What *Should* We Be Worried About?

ALSO BY JOHN BROCKMAN

AS AUTHOR

By the Late John Brockman
37
Afterwords
The Third Culture: Beyond the Scientific Revolution
Digerati

AS EDITOR

About Bateson
Speculations
Doing Science
Ways of Knowing
Creativity
The Greatest Inventions of the Past 2,000 Years
The Next Fifty Years
The New Humanists
Curious Minds
What We Believe But Cannot Prove
My Einstein
Intelligent Thought
What Is Your Dangerous Idea?
What Are You Optimistic About?
What Have You Changed Your Mind About?
This Will Change Everything
Is The Internet Changing the Way You Think?
Culture
The Mind
This Will Make You Smarter
This Explains Everything
Thinking

AS COEDITOR

How Things Are (with Katinka Matson)

NEW YORK • LONDON • TORONTO • SYDNEY • NEW DELHI • AUCKLAND

What *Should* We Be Worried About?

Real Scenarios That Keep Scientists Up at Night

JOHN BROCKMAN

HARPER PERENNIAL

 For information address HarperCollins Publishers, 10 East 53rd Street, New York, NY 10022.

HarperCollins books may be purchased for educational, business, or sales promotional use. For information please email the Special Markets Department at SPsales@harpercollins.com.

FIRST EDITION

Designed by William Ruoto.

Library of Congress Cataloging-in-Publication Data has been applied for.

ISBN 978-0-06-229623-8

14 15 16 17 18 OV/RRD 10 9 8 7 6 5 4 3 2 1

To Daniel Kahneman
who knows from worry

CONTENTS

ACKNOWLEDGMENTS

I wish to thank Peter Hubbard of HarperCollins for his encouragement. I am also indebted to my agent, Max Brockman, who saw the potential for this book, and, as always, to Sara Lippincott for her thoughtful and meticulous editing.

JOHN BROCKMAN

Publisher & Editor, Edge

PREFACE: THE *EDGE* QUESTION

JOHN BROCKMAN

In 1981, I founded the Reality Club, an attempt to gather together those people exploring the themes of the post–Industrial Age. In 1997, the Reality Club went online, rebranded as *Edge*. The ideas presented on *Edge* are speculative; they represent the frontiers in such areas as evolutionary biology, genetics, computer science, neurophysiology, psychology, cosmology, and physics. Emerging out of these contributions is a new natural philosophy, new ways of understanding physical systems, new ways of thinking that call into question many of our basic assumptions.

For each of the anniversary editions of *Edge*, I and a number of *Edge* stalwarts, including Stewart Brand, Kevin Kelly, and George Dyson, get together to plan the annual *Edge* Question—usually one that comes to one or another of us or our correspondents in the middle of the night. It's not easy coming up with a question. (As the late James Lee Byars, my friend and sometime collaborator, used to say: "I can answer the question, but am I bright enough to ask it?") We look for questions that inspire unpredictable answers—that provoke people into thinking thoughts they normally might not have.

The 2013 *Edge* Question: *WHAT* SHOULD *WE BE WORRIED ABOUT?*

We worry because we are built to anticipate the future. Nothing can stop us from worrying, but science can teach us how to worry better, and when to stop worrying. The respondents to this year's question were asked to tell us something that (for scientific reasons) worries them—particularly something that doesn't seem to be on the popular radar yet, and why it should be. Or tell us about something they've stopped worrying about even if others still do, and why it should drop off the radar.

THE REAL RISK FACTORS FOR WAR

STEVEN PINKER
Johnstone Family Professor, Department of Psychology, Harvard University; author, The Better Angels of Our Nature: Why Violence Has Declined

Today the vast majority of the world's people do not have to worry about dying in war. Since 1945, wars between great powers and developed states have essentially vanished, and since 1991 wars in the rest of the world have become fewer and less deadly.

But how long will this trend last? Many people have assured me that it must be a momentary respite, and that a Big One is just around the corner.

Maybe they're right. The world has plenty of unknown unknowns, and perhaps some unfathomable cataclysm will wallop us out of the blue. But since by definition we have no idea what the unknown unknowns are, we can't constructively worry about them.

What, then, about the known unknowns? Are certain risk factors numbering our days of relative peace? In my view, most people are worrying about the wrong ones, or are worrying about them for the wrong reasons.

Resource shortages. Will nations go to war over the last dollop of oil, water, or strategic minerals? It's unlikely. First, resource shortages are self-limiting: As a resource becomes scarcer and thus more expensive, technologies for finding and extracting it improve, or substitutes are found. Also, wars are rarely fought over scarce physical resources (unless you subscribe to the unfalsifiable theory that *all* wars, regardless of stated motives, are really about resources: Vietnam was about tungsten; Iraq was about oil, and so on). Physical resources can be divided or traded, so compromises are always available; not so for psychological motives such as glory, fear, revenge, or ideology.

Climate change. There are many reasons to worry about climate change, but major war is probably not among them. Most studies have failed to find a correlation between environmental degradation and war; environmental crises can cause local skirmishes, but a major war requires a political decision that a war would be advantageous. The 1930s Dust Bowl did not cause an American civil war; when we did have a civil war, its causes were very different.

Drones. The whole point of drones is to minimize loss of life compared to indiscriminate forms of destruction such as artillery, aerial bombardment, tank battles, and search-and-destroy missions, which killed orders of magnitude more people than drone attacks in Afghanistan and Pakistan.

Cyberwarfare. No doubt cyberattacks will continue to be a nuisance, and I'm glad that experts are worrying about them. But the cyber–Pearl Harbor that brings civilization to its knees may be as illusory as the Y2K–bug apocalypse. Should we really expect that the combined efforts of governments, universities, corporations, and programmer networks will be outsmarted for extended periods by some teenagers in Bulgaria? Or by government-sponsored hackers in technologically backward countries? Could they escape detection indefinitely, and would they provoke retaliation for no strategic purpose? And even if they did muck up the Internet for a while, could the damage really compare to being blitzed, firebombed, or nuked?

Nuclear inevitability. It's obviously important to worry about nuclear accidents, terrorism, and proliferation, because of the magnitude of the devastation nuclear weapons could wreak, regardless of the probabilities. But how high are the probabilities? The sixty-eight-year history of non-use of nuclear weapons casts doubt on the common narrative that we are still on the brink of nuclear Armageddon. That narrative requires two extraordinary propositions: (1) That leaders are so spectacularly irrational, reckless, and suicidal that they have kept the world in jeopardy of mass annihilation, and (2) we have enjoyed a spectacularly improbable run of good luck. Perhaps. But instead of believing in two riveting and unlikely propositions, perhaps we should believe in one boring and likely one: that world leaders, although stupid and short-sighted, are not *that* stupid and short-sighted and have taken steps to minimize the chance of nuclear war, which is why nuclear war has not taken place. As for nuclear terrorism, though there was a window of vulnerability for theft of weapons and fissile material after the fall of the Soviet Union, most nuclear security experts believe it has shrunk and will soon be closed (see John Mueller's *Atomic Obsession*).

What the misleading risk factors have in common is that they contain the cognitive triggers of fear documented by Slovic, Kahneman, and Tversky: They are vivid, novel, undetectable, uncontrollable, catastrophic, and involuntarily imposed on their victims.

In my view, there *are* threats to peace that we should worry about, but the real risk factors—the ones that actually caused catastrophic wars, such as the World Wars, wars of religion, and the major civil wars—don't press the buttons of our lurid imaginations:

Narcissistic leaders. The ultimate weapon of mass destruction is a state. When a state is taken over by a leader with the classic triad of narcissistic symptoms—grandiosity, need for admiration, and lack of empathy—the result can be imperial adventures with enormous human costs.

Groupism. The ideal of human rights—that the ultimate moral good is the flourishing of individual people, while groups are social constructions designed to further that good—is surprisingly recent and unnatural. People, at least in public, are apt to argue that the ultimate moral good is the glory of the group—the tribe, religion, nation, class, or race—and that individuals are expendable, like the cells of a body.

Perfect justice. Every group has suffered depredations and humiliations in its past. When groupism combines with the thirst for revenge, a group may feel justified in exacting damage on some other group, inflamed by a moralistic certitude that makes compromise tantamount to treason.

Utopian ideologies. If you have a religious or political vision of a world that will be infinitely good forever, any amount of violence is justified to bring about that world, and anyone standing in its way is infinitely evil and deserving of unlimited punishment.

Warfare as a normal or necessary tactic. Clausewitz characterized war as "the continuation of policy by other means." Many political and religious ideologies go a step further and consider violent struggle to be the driver of dialectical progress, revolutionary liberation, or the realization of a messianic age.

THE RELATIVE PEACE we have enjoyed since 1945 is a gift of values and institutions that militate against these risks. Democracy selects for responsible stewards rather than charismatic despots. The ideal of human rights protects people from being treated as cannon fodder, collateral damage, or eggs to be broken for a revolutionary omelet. The maximization of peace and prosperity has been elevated over the rectification of historic injustices or the implementation of utopian fantasies. Conquest is stigmatized as "aggression" and becomes a taboo rather than a natural aspiration of nations or an everyday instrument of policy.

None of these protections is natural or permanent, and the possibility of their collapsing is what makes me worry. Perhaps some charismatic politician is working his way up the Chinese nomenklatura and dreams of overturning the intolerable insult of Taiwan once and for all. Perhaps an aging Putin will seek historical immortality and restore Russian greatness by swallowing a former Soviet republic or two. Perhaps a utopian ideology is fermenting in the mind of a cunning fanatic somewhere who will take over a major country and try to impose it elsewhere.

It's natural to worry about physical stuff like weaponry and resources. What we should really worry about is psychological stuff like ideologies and norms. As the UNESCO slogan puts it, "Since wars begin in the minds of men, it is in the minds of men that the defenses of peace must be constructed."

MADness

VERNOR VINGE
Mathematician; computer scientist; Hugo Award–winning novelist, A Fire upon the Deep, Rainbows End

There are many things we know to worry about. Some are very likely events but by themselves not existential threats to civilization. Others could easily destroy civilization and even life on Earth—but the chances of such disasters occurring in the near historical future seem to be vanishingly small.

There is a known possibility that stands out for being both likely in the next few decades and capable of destroying our civilization. It's prosaic and banal, something dismissed by many as a danger that the 20th century confronted and definitively rejected: That is war between great nations, especially when fought under a doctrine of Mutually Assured Destruction (MAD).

Arguments against the plausibility of MAD warfare are especially believable these days: MAD war benefits no one. Twentieth-century U.S.A. and U.S.S.R., even in the depths of the MAD years, were sincerely desperate to avoid tipping over into MAD warfare. That sincerity is a big reason why humanity got through the century without general nuclear war.

Unfortunately, the 20th century is our only test case, and the MAD warfare threat has characteristics that made surviving the 20th century more a matter of luck than wisdom.

MAD involves very long time scales and very short ones. At the long end, the threat is driven by social and geopolitical issues in much the same way as with unintended wars of the past. At the other extreme, MAD involves complex automation controlling large systems, operating faster than any real-time human response, much less careful judgment.

Breakers (vandals, griefers) have more leverage than Makers (builders, creators), even though the Makers far outnumber the Breakers. This is the source of some of our greatest fears about technology—that if weapons of mass destruction are cheap enough, then the relatively small percentage of Breakers will be sufficient to destroy civilization. If that possibility is scary, then the MAD threat should be terrifying. For with MAD planning, it is hundreds of thousands of creative and ingenious people in the most powerful societies—many of the best of the Makers, powered by the riches of the planet—who work to create a mutually unsurvivable outcome! In the most extreme case, the resulting weapon systems must function on the shortest of time scales, thus moving the threat into the realm of thermodynamic inevitability.

For the time (decades?) in which we and our interests are undefendable and still confined to a volume smaller than the scope of our weapons, the threat of MAD warfare will be the winner in rankings of likely destructiveness.

There's a lot we can do to mitigate the threat of MADness:

A resurrection of full-blown MAD planning will probably be visible to the general public. We should resist arguments that MAD doctrine is a safe strategy with regard to weapons of mass destruction.

We should study the dynamics of the beginning of unintended wars of the past—in particular, World War I. There are plenty of similarities between our time and the first few years of the last century. We have much optimism, the feeling that our era is different. And what about entangling alliances? Are there small players with the ability to bring heavyweights into the action?

How does the possibility of *n*-way MADness affect these risks?

With all the things we have to worry about, there is also an overwhelmingly positive counterweight: billions of good, smart people and the databases and networks that now empower them. This is an intellectual force that trumps all institutions of the past. Humanity plus its automation is quite capable of anticipating and countering myriad possible calamities. If we can avoid blowing ourselves up, we will have time to create things so marvelous that their *upside* is (worrisomely!) beyond imagination.

WE ARE IN DENIAL ABOUT CATASTROPHIC RISKS

MARTIN REES

Astronomer Royal; former president, the Royal Society; emeritus professor of cosmology & astrophysics, University of Cambridge; author, From Here to Infinity: A Vision for the Future of Science

Those of us fortunate enough to live in the developed world fret too much about minor hazards of everyday life: improbable air crashes, carcinogens in food, and so forth. But we are less secure than we think. We should worry far more about scenarios that have thankfully not yet happened—but which, if they occurred, could cause such worldwide devastation that even once would be too often.

Much has been written about possible ecological shocks triggered by the collective impact on the biosphere of a growing and more demanding world population, and about the social and political tensions stemming from scarcity of resources or climate change. But even more worrying are the downsides of powerful new technologies: cyber-, bio-, and nano-. We're entering an era when a few individuals could, via error or terror, trigger a societal breakdown with such extreme suddenness that palliative government actions would be overwhelmed.

Some would dismiss these concerns as an exaggerated jeremiad: After all, human societies have survived for millennia despite storms, earthquakes, and pestilence. But these human-induced threats are different: They are newly emergent, so we have a limited time base for exposure to them and can't be so sanguine that we would survive them for long, nor about the ability of governments to cope if disaster strikes. And of course we have zero grounds for confidence that we can survive the worst that even more powerful future technologies could do.

The "anthropocene" era, when the main global threats come from humans and not from nature, began with the mass deployment of thermonuclear weapons. Throughout the cold war, there were several occasions when the superpowers could have stumbled toward nuclear Armageddon through muddle or miscalculation. Those who lived anxiously through the Cuban missile crisis would have been not merely anxious but paralytically scared had they realized just how close the world then was to catastrophe. Only later did we learn that President Kennedy assessed the odds of nuclear war, at one stage, as "somewhere between one out of three and even." And only when he was long retired did Robert MacNamara state frankly that "[w]e came within a hair's breadth of nuclear war without realizing it. It's no credit to us that we escaped—Khrushchev and Kennedy were lucky as well as wise."

It is now conventionally asserted that nuclear deterrence worked. In a sense, it did. But that doesn't mean it was a wise policy. If you play Russian roulette with one or two bullets in the barrel, you are more likely to survive than not, but the stakes would need to be astonishingly high—or the value you place on your life inordinately low—for this to seem a wise gamble.

But we were dragooned into just such a gamble throughout the cold war era. It would be interesting to know what level of risk other leaders thought they were exposing us to, and what odds most European citizens would have accepted, if they'd been asked to give informed consent. For my part, I would not have chosen to risk a one in three—or even one in six—chance of a disaster that would have killed hundreds of millions and shattered the physical fabric of all our cities, even if the alternative were a certainty of a Soviet invasion of Western Europe. And of course the devastating consequences of thermonuclear war would have spread far beyond the countries that faced a direct threat.

The threat of global annihilation involving tens of thousands of H-bombs is thankfully in abeyance—even though there is now more reason to worry that smaller nuclear arsenals might be used in a regional context, or even by terrorists. But when we recall the geopolitical convulsions of the last century—two world wars, the rise and fall of the Soviet Union, and so forth—we can't rule out, later in the present century, a drastic global realignment leading to a standoff between new superpowers. So a new generation may face its own "Cuba"—and one that could be handled less well or less luckily than the Cuban missile crisis was.

We will always have to worry about thermonuclear weapons. But a new trigger for societal breakdown will be the environmental stresses consequent on climate change. Many still hope that our civilization can segue toward a low-carbon future without trauma and disaster. My pessimistic guess, however, is that global annual CO_2 emissions won't be turned around in the next twenty years. But by then we'll know—perhaps from advanced computer modeling but also from how much global temperatures have actually risen by then—whether or not the feedback from water vapor and clouds strongly amplifies the effect of CO_2 itself in creating a greenhouse effect.

If these feedbacks are indeed important, and the world consequently seems on a rapidly warming trajectory because international efforts to reduce emission haven't been successful, there may be a pressure for "panic measures." These would have to involve a "Plan B"—being fatalistic about continuing dependence on fossil fuels but combating its effects by some form of geoengineering.

That would be a political nightmare: Not all nations would want to adjust the thermostat the same way, and the science would still not be reliable enough to predict what would actually happen. Even worse, techniques such as injecting dust into the stratosphere or "seeding" the oceans may become cheap enough that plutocratic individuals could finance and implement them. This is a recipe for dangerous and possible runaway unintended consequences, especially if some want a warmer Arctic whereas others want to avoid further warming of the land at lower latitudes.

Nuclear weapons are the worst downside of 20th-century science. But there are novel concerns stemming from the effects of fast-developing 21st-century technologies. Our interconnected world depends on elaborate networks: electric power grids, air-traffic control, international finance, just-in-time delivery, and so forth. Unless these are highly resilient, their manifest benefits could be outweighed by catastrophic (albeit rare) breakdowns cascading through the system.

Moreover, a contagion of social and economic breakdown would spread worldwide via computer networks and "digital wildfire"—literally at the speed of light. The threat is terror as well as error. Concern about cyberattack, by criminals or by hostile nations, is rising sharply. Synthetic biology, likewise, offers huge potential for medicine and agriculture—but it could facilitate bioterror.

It is hard to make a clandestine H-bomb, but millions will have the capability and resources to misuse these "dual use" technologies. Freeman Dyson looks toward an era when children can design and create new organisms just as routinely as he, when young, played with a chemistry set. Were this to happen, our ecology (and even our species) would surely not survive unscathed for long. And should we worry about another sci-fi scenario—that a network of computers could develop a mind of its own and threaten us all?

In a media landscape oversaturated with sensational science stories, "end of the world" Hollywood productions, and Mayan apocalypse warnings, it may be hard to persuade the wide public that there are indeed things to worry about that could arise as unexpectedly as the 2008 financial crisis and have far greater impact. I'm worried that by 2050 desperate efforts to minimize or cope with a cluster of risks with low probability but catastrophic conseqences may dominate the political agenda.

LIVING WITHOUT THE INTERNET FOR A COUPLE OF WEEKS

DANIEL C. DENNETT
Philosopher, University Professor, codirector, Center for Cognitive Studies, Tufts University; author, Intuition Pumps and Other Tools for Thinking

In the early 1980s, I was worried that the computer revolution was going to reinforce and amplify the divide between the (well-to-do, Western) technocrats and those around the world who couldn't afford computers and similar high-tech gadgetry. I dreaded a particularly malignant sorting of the haves and have nots, with the rich getting ever richer and the poor being ever more robbed of political and economic power by their lack of access to the new information technology. I started devoting some serious time and effort to raising the alarm about this, and trying to think of programs that would forestall or alleviate it, but before I'd managed to make any significant progress the issue was happily swept out of my hands by the creation of the Internet. I was an Arpanet user, but that didn't help me anticipate what was coming.

We've certainly seen a lot of rich technocrats getting richer, but we've also seen the most profoundly democratizing and leveling spread of technology in history. Cell phones and laptops, and now smartphones and tablets, put worldwide connectivity in the hands of billions, adding to the inexpensive transistor radios and television sets that led the way. The planet has become informationally transparent in a way nobody imagined only forty years ago.

This is wonderful, mostly. Religious institutions that could always rely in the past on the relative ignorance of their flock must now revise their proselytizing and indoctrinating policies or risk extinction. Dictators face the dire choice between maximal suppression—turning their nations into prisons—or tolerating an informed and well-connected opposition. Knowledge really *is* power, as people are coming to realize all over the world.

This leveling does give us something new to worry about, however. We have become so dependent on this technology that we have created a shocking new vulnerability. We really don't have to worry much about an impoverished teenager making a nuclear weapon in his slum; it would cost millions of dollars and be hard to do inconspicuously, given the exotic materials required. But such a teenager with a laptop and an Internet connection can explore the world's electronic weak spots for hours every day, almost undetectably at almost no cost and very slight risk of being caught and punished. Yes, the Internet is brilliantly designed to be so decentralized and redundant that it's almost invulnerable, but robust as it is, it isn't perfect.

Goliath hasn't been knocked out yet, but thousands of Davids are busily learning what they need to know to contrive a trick that will even the playing field with a vengeance. They may not have much money, but we won't have any either, if the Internet goes down. I think our choice is simple: We can wait for them to annihilate what we have, which is becoming more likely every day, or we can begin thinking about how to share what we have with them.

In the meantime, it would be prudent to start brainstorming about how to keep panic at bay if a long-term disruption of large parts of the Internet were to occur. Will hospitals and fire stations (and supermarkets and gas stations and pharmacies) keep functioning, and how will people be able to get information they trust? Cruise ships oblige their paying customers to walk through a lifeboat drill the first day at sea, and while it isn't a popular part of the cruise, people are wise to comply. Panic can be contagious, and when that happens, people make crazy and regrettable decisions. As long as we insist on living in the fast lane, we should learn how to get on and off without creating mayhem.

Perhaps we should design and institute nationwide lifeboat drills to raise consciousness about what it would be like to have to cope with a long-term Internet blackout. When I try to imagine what the major problems would be and how they could be coped with, I find I have scant confidence in my hunches. Are there any experts on this topic?

SAFE MODE FOR THE INTERNET

GEORGE DYSON
Science historian; author, Turing's Cathedral: The Origins of the Digital Universe

Sooner or later—by intent or by accident—we will face a catastrophic breakdown of the Internet. Yet we have no Plan B in place to reboot a rudimentary, low-bandwidth emergency communication network if the high-bandwidth system we've come to depend on fails.

In the event of a major network disruption, most of us will have no idea what to do except to try and check the Internet for advice. As the system begins to recover, the resulting overload may bring that recovery to a halt.

The ancestor of the Internet was the store-and-forward punched-paper-tape telegraph network. This low-bandwidth, high-latency system was sufficient to convey important messages, like "Send ammunition" or "Arriving New York Dec. 12. Much love. Stop."

We need a low-bandwidth, high-latency store-and-forward message system that can run in emergency mode on an ad-hoc network assembled from mobile phones and laptop computers even if the main networks fail. We should keep this system on standby and periodically exercise it, along with a network of volunteers trained in network first aid the way we train lifeguards and babysitters in CPR. These first responders, like the amateur radio operators who restore communications after natural disasters, would prioritize essential communications, begin the process of recovery, and relay instructions as to what to do next.

Most computers—from your car's engine controller to your desktop—can be rebooted into safe mode to get you home. But no safe mode for the Internet? We should be worried about that.

THE FRAGILITY OF COMPLEX SYSTEMS

RANDOLPH NESSE
Professor of psychiatry & psychology, University of Michigan; coauthor (with George C. Williams), Why We Get Sick

On the morning of August 31, 1859, the sun ejected a giant burst of charged particles. They hit Earth eighteen hours later, creating auroras so bright that at 1:00 A.M. birds sang and people thought morning had dawned. Currents induced in telegraph wires prevented transmission, and sparks from the wires set papers aflame. According to data from ice cores, solar ejections this intense occur about every 500 years. A 2008 National Academy of Sciences report concluded that a similar event now would cause "extensive social and economic disruptions." Power outages would last for months, and there would be no GPS navigation, cell phone communication, or air travel.

Geomagnetic storms sound like a pretty serious threat. But I am far less concerned about them than I am about the effects of many possible events on the complex systems we have become dependent on. Any number of events that once would have been manageable now will have catastrophic effects. Complex systems like the markets, transportation, and the Internet seem stable, but their complexity makes them inherently fragile. Because they are efficient, massive complex systems grow like weeds, displacing slow markets, small farmers, slow communication media, and local information-processing systems. When they work, they are wonderful, but when they fail, we will wonder why we did not recognize the dangers of depending on them.

It would not take a geomagnetic storm to stop trucks and planes from transporting the goods that make modern life possible; an epidemic or bioterrorist attack would be sufficient. Even a few decades ago, food was produced close to population centers. Now world distribution networks prevent famine nearly everywhere—and make mass starvation more likely if they are disrupted suddenly. Accurate GPS has been available to civilians for less than twenty years. When it fails, commuters will only be inconvenienced, but most air and water transport will stop. The Internet was designed to survive all manner of attacks, but our reckless dependency on it is nonetheless astounding. When it fails, factories and power stations will shut down, air and train travel will stop, hospitals and schools will be paralyzed, and most commerce will cease. What will happen when people cannot buy groceries? "Social chaos" is a pallid phrase for the likely scenarios.

Modern markets exemplify the dangers of relying on complex systems. Economic chaos from the failures of massively leveraged bets is predictable. That governments have been unable to establish controls is astonishing, given that the world economic system came within days of collapse just five years ago. Complex trading systems fail for reasons that are hard to grasp, even by investigations after the fact. The Flash Crash of May 6, 2010, wiped out over a trillion dollars of value in minutes, thanks to high-frequency trading algorithms interacting with one another in unpredictable ways. You might think this would have resulted in regulations to prevent any possibility of reccurrence, but mini-flash crashes continue and the larger system remains vulnerable.

These are examples because they have already happened. The larger dangers come from the hidden fragility of complex systems. James Crutchfield, of the Complexity Sciences Center · UC Davis, has written clearly about the risks, but as far as I can tell few are paying attention. We should. Protecting us from catastrophes caused by our dependency on fragile complex systems is something governments can and should do. We need to shift our focus from this or that threat to the vulnerabilities of modern complex systems to any number of threats. Our body politic is like an immune compromised patient, vulnerable to collapse from numerous agents. Instead of just studying the threats, we need scientists to study the various ways that complex systems fail, how to identify those that make us most vulnerable, and what actions can prevent otherwise inevitable catastrophes.

A SYNTHETIC WORLD

SEIRIAN SUMNER

Senior lecturer, School of Biological Sciences, University of Bristol

Synthetic biology is Legoland for natural scientists. We take nature's building blocks apart and piece them back together again in a way that suits us better. We can combine genetic functions to reprogram new biological pathways with predictable behaviors. We spent the last decade imagining how this will improve society and the environment. We are now realizing these dreams. We can make yogurt that mops up cholera; we can manufacture yeast to power our cars; we can engineer microorganisms to clean up our environment. Soon we'll be using living organisms to mimic electrical engineering solutions—biocomputers programmed to follow logic gates just as computers do. We will have materials stronger than steel, made from animal products. Could this be the end of landfill? There's no doubt that synthetic biology will revolutionize our lives in the 21st century.

I worry about where synthetic biology is going next, and specifically what happens when it gets out of the lab into the natural world and the public domain.

Biological engineering started outside the lab; we've been modifying plants and animals since the advent of agriculture, about 12,000 years ago, through breeding and artificial selection for domestication. We've ensnared yeast and bacteria to make beer, wine, and cheese; we've tamed wolves to be man's best friend; we've cajoled grass into being a high nutrient source. Synthetic biology is a new packaging that describes how we've got an awful lot better at manipulating natural systems to suit our whims. A "plug and play" approach is being developed (e.g., BioBricks) to facilitate manipulations at the molecular level. In the future, tried and tested genetic modules may be slotted together by nonexperts to create their own bioengineered product. Our children's children could be getting Bio-Lego for Christmas to build their own synthetic pets!

Synthetic biology has tremendous commercial potential (beyond the Lego) and is estimated to be worth over $10 billion by 2016. Currently, progress is focused on small things, like individual gene networks or microorganisms. But there is potential, too, for the larger, more charismatic organisms—specifically, the fluffy or endangered ones. These species capture the interests of the public, business, and entrepreneurs. This is what I am worried about.

We can make a new whole organism from a single stem cell (e.g., Dolly & Co.). We can uncover the genome sequence, complete with epigenetic programming instructions, for practically any extant organism within a few weeks. With this toolkit, we could potentially re-create any living organism on the planet; animal populations on the brink of extinction could be restocked with better, hardier forms. We are a stone's throw away from re-creating extinct organisms.

The woolly mammoth genome was sequenced in 2008, and Japanese researchers are reputedly cloning it now, using extant elephant relatives as surrogate mothers. Synthetic biology makes resurrecting extinct animals much more achievable, because any missing genomic information can be replaced with a plug-and-play genetic module. A contained collection of resurrected animals is certainly a Wow-factor, and it might help uncover their secret lives and explain why they went extinct. But as Hollywood tells us, even a Jurassic Park cannot be contained for long.

There are already attempts to re-create ancient ecosystems

through the reintroduction of the descendants of extinct megafauna (e.g., Pleistocene Park, in Russia), and synthetic woolly mammoths may complete the set. Could synthetic biology be used to resurrect species that "fit better" or present less of a threat to humans? A friendly mammoth perhaps? Extinct, extant, friendly, or fierce, I worry about the consequences of biosynthetic aliens being introduced into a naïve and vulnerable environment, becoming invasive, and devastating native ecosystems. I worry that if we can re-create any animal, why should we bother conserving any in the first place?

Synthetic biology is currently tightly regulated, along the same lines as genetically modified organisms (GMOs). But when biosynthetic products overflow into the natural world, it will be harder to keep control. Let's look at this from the molecular level, which arguably we have more control over than the organism level or the ecosystem level. We can shuffle genes or whole genomes to create something that nature did not get around to creating. But a biological unit does not exist in isolation: Genes, protein complexes, and cells all function in modules—a composite of units, finely tuned by evolution in a changeable environment to work together.

Modules may be swapped around, allowing plasticity in a system. But there are rules to "rewiring." Synthetic biology relies on a good understanding of these rules. Do we really understand the molecular rules enough to risk releasing our synthetic creations into natural ecosystems? We barely understand the epigenetic processes that regulate cell differentiation in model organisms in controlled lab conditions. How do we deal with the epigenome in a synthetic genome, especially one destined to exist in an environment very different from its original one 10,000 years ago?

Ecology is the Play-Doh of evolution: Ecosystem components get pushed and pulled, changing form, function, and relationships. We might be able to create a biounit that looks perfect and performs perfectly in the lab, but we cannot control how ecology and evolution might rewire our synthetic unit in an ecosystem, nor can we predict how that synthetic unit might rewire the ecosystem and its inhabitants. Molecular control mechanisms are engineered into the microorganisms we use to clean up toxic spills in the environment, preventing them from evolving and spreading. Can we put a "Stop evolving" switch into a more complex organism? How do we know that it won't evolve around such a switch? And what happens if (when) such organisms interbreed with native species? What the disruption of the engineered modules, or their transfer to other organisms, might lead to is unimaginable.

To sum up, I worry about the natural world becoming naturally unnatural.

WHAT IS CONSCIOUS?

TIMO HANNAY
Managing director, Digital Science; former director, Nature.com; co-organizer, Sci Foo

In an episode of the 1980s British TV series *Tales of the Unexpected*, an amateur botanist discovers that plants emit agonized screams when pruned, though at frequencies well beyond human hearing. So overcome is he with sympathy for this suffering vegetation, and so apparently bizarre his demands that a local doctor give medical attention to his trees, that he is quickly packed off to an asylum.

A preposterous flight of fancy from Roald Dahl's ever fertile imagination? No doubt. But this conceit also raises a profoundly serious point: We have next to no idea which things in the world around us are conscious and which are not.

This might seem like an abstract philosophical issue, but on the contrary, consciousness is the substrate of all suffering and pleasure and thus the mediator of everything truly important to us. If there were no subjective experience, there would be no such things as kindness, love, or joy. However sublime our universe, it would be inconsequential without a consciousness to perceive it. True, in a world with no subjective experience there would also be no cruelty, pain, or worrying (including that of the kind I am doing now). But this is precisely the point: How are we to maximize happiness and minimize suffering if we do not reliably know where and when they can exist?

Our usual rule of thumb for the presence of consciousness is to assess, based on superficial cues, how similar to ourselves something appears to be. Thus a dog is more conscious than a duck, which in turn is more conscious than a daffodil. But our intuitions about so many things—from the motions of celestial objects to the likelihood of winning the lottery—are so often wrong that we are foolish to rely on them for something as important as the ultimate source of all joy and strife.

Famously, and ironically, the only thing of which we can be truly certain is the existence of our own subjective experience, and we see the physical world only through this dark glass. Yet the scientific method has proved a remarkable tool for clarifying our view and enabling us to develop an elaborate, apparently objective consensus about how the world works. Unfortunately, having provided us with an escape route from our own subjectivity, science leaves us almost completely impotent to probe the nature and origins of subjective experience itself. The truth is that we have no idea what things have consciousness, where it comes from, or even what it is. All we really know is how it feels.

That may sound like an odd statement to come from a former neuroscientist. Certainly we understand an impressive amount about how the brain functions as a physical system, and also about the ways in which different brain states correspond to various reported subjective experiences. But this is a long way from understanding consciousness well enough to be able to do what really matters: determine with reasonable certainty what does and does not possess it, and to what degree.

This is a fabulously hard task. Daniel Dennett's outstanding book *Consciousness Explained*, for all its considerable eloquence and erudition, falls well short of the claim of its title. Indeed, we have tied ourselves in such intellectual knots over consciousness that it is hard to discern more than incremental progress since Descartes.

For an example of the difficulties involved, consider John Searle's celebrated Chinese Room thought experiment, which purports to show—contrary to Alan Turing's claims—that input-output characteristics alone are insufficient to determine the existence of a conscious mind. Intuitively this conclusion seems right: A sleeping person, immobile and inattentive, might nevertheless be experiencing vivid dreams. Conversely, I can drive a familiar route without forming any conscious record of the journey that took me to my destination. But the Chinese Room does nothing to prove this thesis, for it is a thought experiment, and the trouble with thought experiments is that the researcher chooses not only the experimental conditions but also the results. This makes them useful for testing the internal consistency of ideas but almost useless for probing mysterious, apparently emergent phenomena like consciousness. (To see this, carry out the same thought experiment on the 1.4-kilogram lump of electrophysiological goo called the human brain, and if you're being consistent you'll get the same result: There appears to be no conscious understanding anywhere inside.)

Over the last decade or two, neuroscientists have at last shed their qualms about investigating the mysterious and transcendent phenomenon of consciousness, with some interesting results. Our best guess these days is that consciousness arises when certain types and quantities of information are integrated in the brain in certain ways. But this is still very hand-wavy, not only because the parameters are so ill-defined but also because we don't even know exactly what we mean by "information." After all, every physical system contains information of one sort or another, and "computes" its own behavior, so the brain is far from unique in this respect.

What's more, even if we understood exactly what kind of information has to be brought together in precisely what combinations and quantities to ignite a spark of consciousness, we'd still be in the dark as to whether this is an emergent or fundamental phenomenon and why this physical universe even allows subjective experience to exist at all. Those mysteries remain as dark to us as the nature of existence itself.

Toward the end of 2012, doctors picked up the first message (via a brain scan) from a patient in a persistent vegetative state. But if input-output characteristics can't be trusted—and they probably can't—then we really have no way of confirming whether this represented a conscious act or an insensible physical response. Similarly, what of patients under anesthetic? For all we know, they may be in agony during their operations even if they have no memory of it afterward. And so on and on, from human embryos to birds, insects, and, yes, even plants. Not to mention all those computers we're bringing into existence: Might they have an inner life, too?

It is possible that we are rare, fleeting specks of awareness in an unfeeling cosmic desert, the only witnesses to its wonder. It is also possible that we are living in a universal sea of sentience, surrounded by ecstasy and strife that is open to our influence. Sensible beings that we are, both possibilities should worry us.

WILL THERE BE A SINGULARITY WITHIN OUR LIFETIME?

MAX TEGMARK

Physicist, MIT; researcher, precision cosmology; scientific director, FQXi (Foundational Questions Institute)

Although life as we know it gets a lot of flak, I worry that we don't appreciate it enough and are too complacent about losing it.

As our Spaceship Earth blazes though cold and barren space, it both sustains and protects us. It's stocked with major but limited supplies of water, food, and fuel. Its atmosphere keeps us warm and shielded from the sun's harmful ultraviolet rays. Its magnetic field shelters us from lethal cosmic rays. Surely any responsible spaceship captain would make it a top priority to safeguard his craft's future existence by avoiding asteroid collisions, onboard explosions, overheating, ultraviolet-shield destruction, and premature depletion of ship supplies. Yet our spaceship crew hasn't made any of these issues a top priority, devoting (by my estimate) less than a millionth of its resources to them. In fact, our spaceship doesn't even have a captain!

Many have blamed this dismal performance on life as we know it, arguing that since our environment is changing, we humans need to change with it: We need to be technologically enhanced, perhaps with smartphones, smartglasses, brain implants, and ultimately by merging with superintelligent computers. Does the idea of life as we know it getting replaced by more advanced life sound appealing or appalling to you? That probably depends on the circumstances—and in particular on whether you view the future beings as our descendants or our conquerors.

If parents have a child who's smarter than they are, who learns from them and then goes out and accomplishes what they could only dream of, they'll probably feel happy and proud, even if they know they can't live to see it all. Parents of a highly intelligent mass murderer feel differently. We might feel that we have a similar parent-child relationship with future AIs, regarding them as the heirs of our values. It will therefore make a huge difference whether or not future advanced life retains our most cherished goals.

Another key factor is whether the transition is gradual or abrupt. I suspect that few are disturbed by the prospects of humankind gradually evolving, over thousands of years, to become more intelligent and better adapted to our changing environment, perhaps also modifying its physical appearance in the process. On the other hand, many parents would feel ambivalent about having their dream child if they knew it would cost them their lives. If advanced future technology doesn't replace us abruptly but rather upgrades and enhances us gradually, eventually merging with us, then this might provide both the goal retention and the gradualism required for us to view future technological life-forms as our descendants.

So what will actually happen? This is something we should be really worried about. The Industrial Revolution has brought us machines that are stronger than we are. The Information Revolution has brought us machines that are smarter than we are in certain limited ways, beating us in chess in 2006, on the quiz show *Jeopardy!* in 2011, and at driving in 2012, when a computer was licensed to drive cars in Nevada after being judged safer than a human. Will computers eventually beat us at all tasks, developing superhuman intelligence?

I have little doubt that this can happen: Our brains are a

bunch of particles obeying the laws of physics, and there's no physical law precluding particles from being arranged in ways that can perform even more advanced computations.

But will it happen anytime soon? Many experts are skeptical, while others, such as Ray Kurzweil, predict it will happen by 2030. What I think is quite clear, however, is that if it happens, the effects will be explosive. As the late Oxford mathematician Irving J. Good realized in 1965 ("Speculations Concerning the First Ultraintelligent Machine"), machines with superhuman intelligence could rapidly design even better machines. In 1993, mathematician and science-fiction author Vernor Vinge called the resulting intelligence explosion "The Singularity," arguing that it was a point beyond which it was impossible for us to make reliable predictions. After this, life on Earth would never be the same, either objectively or subjectively.

Objectively, whoever or whatever controls this technology would rapidly become the world's wealthiest and most powerful entity, outsmarting all financial markets, outinventing and outpatenting all human researchers, and outmanipulating all human leaders. Even if we humans nominally merge with such machines, we might have no guarantees about the ultimate outcome, making it feel less like a merger and more like a hostile corporate takeover.

Subjectively, these machines wouldn't feel as we do. Would they feel anything at all? I believe that consciousness is the way information feels when being processed. I therefore think it's likely that they, too, would feel self-aware and should be viewed not as mere lifeless machines but as conscious beings like us—but with a consciousness that subjectively feels quite different from ours.

For example, they would probably lack our human fear of death. As long as they've backed themselves up, all they stand to lose are the memories they've accumulated since their latest backup. The ability to readily copy information and software between AIs would probably reduce the strong sense of individuality so characteristic of human consciousness: There would be less of a distinction between you and me if we could trivially share and copy all our memories and abilities. So a group of nearby AIs may feel more like a single organism with a hive mind.

In summary, will there be a Singularity within our lifetime? And is this something we should work for or against? On the one hand, it might solve most of our problems, even mortality. It could also open up space, the final frontier. Unshackled by the limitations of our human bodies, such advanced life could rise up and eventually make much of our observable universe come alive. On the other hand, it could destroy life as we know it and everything we care about.

We're nowhere near consensus on either of these two questions, but that doesn't mean it's rational for us to do nothing about the issue. It could be the best or worst thing ever to happen to life as we know it, so if there's even a 1-percent chance that there will be a Singularity in our lifetime, a reasonable precaution would be to spend at least 1 percent of our GDP studying the issue and deciding what to do about it. Yet we largely ignore it and are curiously complacent about life as we know it getting transformed. What we should be worried about is that we're not worried.

"THE SINGULARITY": THERE'S NO THERE THERE

BRUCE STERLING
Futurist, science fiction author, journalist, critic; author, Love Is Strange (A Paranormal Romance)

Twenty years have passed since Vernor Vinge wrote his remarkably interesting essay about the Singularity.

This aging sci-fi notion has lost its conceptual teeth. Plus, its chief evangelist, visionary Ray Kurzweil, recently got a straight engineering job with Google. Despite its weird fondness for AR goggles and self-driving cars, Google is not going to finance any eschatological cataclysm in which superhuman intelligence abruptly ends the human era. Google is a firmly commercial enterprise.

It's just not happening. All the symptoms are absent. Computer hardware is not accelerating on any exponential runway beyond all hope of control. We're no closer to self-aware machines than we were in the remote 1960s. Modern wireless devices in a modern cloud are an entirely different cyberparadigm than imaginary 1990s "minds on nonbiological substrates" that might allegedly have the "computational power of a human brain." A Singularity has no business model, no major power group in our society is interested in provoking one, nobody who matters sees any reason to create one, there's no there there.

So, as a pope once remarked, "Be not afraid." We're getting what Vinge predicted would happen without a Singularity, which is "a glut of technical riches never properly absorbed." There's all kinds of mayhem in that junkyard, but the AI Rapture isn't lurking in there. It's no more to be fretted about than a landing of Martian tripods.

CAPTURE

CHARLES SEIFE
Professor of journalism, NYU; former staff writer, Science*;*
author, Proofiness: The Dark Arts of Mathematical Deception

On April 5, 2010, deep in the Upper Big Branch mine in West Virginia, a spark ignited a huge explosion that rumbled through the tunnels and killed twenty-nine miners—the worst mining disaster in the United States in forty years. Two weeks later, the Deepwater Horizon, a drilling rig in the Gulf of Mexico, went up in flames, killing eleven workers and creating the biggest oil spill in history. Though these two disasters seem completely unrelated, they had the same underlying cause: capture.

Federal agencies that regulate industry are supposed to prevent such disasters. Agencies like the Mine Safety and Health Administration (which sets the rules for mines) and the Minerals Management Service (which set the rules for offshore drilling) are supposed to constrain businesses—and to act as watchdogs—to force everyone to play by the rules. That's the ideal, anyhow. The reality is a bit messier. More often than not, the agencies are reluctant to enforce the regulations they create. When a business gets caught breaking the rules, the regulatory agencies tend to impose penalties amounting to no more than a slap on the wrist. Companies like Massey Energy (which ran Upper Big Branch) and BP (which ran the Deepwater Horizon) flout the rules, and when disaster strikes, everybody wonders why regulators failed to take action despite numerous warning signs and repeated violations of regulations.

In the 1970s, economists, led by future Nobel laureate George

Stigler, began to realize that this was the rule, not the exception. Over time, regulatory agencies are systematically drained of their ability to check the power of industry. Even more striking, they're gradually drawn into the orbit of the businesses they're charged with regulating. Instead of acting in the public interest, the regulators wind up as tools of the industry they're supposed to keep watch over. This process, known as "regulatory capture," turns regulators from watchdogs into lapdogs.

You don't have to look far to see regulatory capture in action. Securities and Exchange Commission officials are often accused of ignoring warnings about fraud, stifling investigations, even helping miscreants avoid paying big fines or going to jail. Look at the Nuclear Regulatory Commission's enforcement reports to see how capable it is of preventing energy companies from violating nuclear power plant safety rules again and again. Regulatory capture isn't limited to the U.S. What caused the Fukushima disaster? Ultimately it was a "breakdown of the regulatory system" caused by "reversal of the positions between the regulator and the regulated," at least according to a report prepared by the Japanese parliament. The regulator had become the regulated.

Regulatory capture is just a small part of the story. In my own profession, journalism, we like to think of ourselves as watchdogs, fierce defenders of the public good. But we, too, are being captured by the industries we're supposed to keep watch on. There's journalistic capture just as there's regulatory capture. It's most marked in fields such as tech reporting, business reporting, White House reporting—fields where you're afraid of losing access to your subjects, where you depend on the industry to feed you stories, where your advertising revenue comes from the very people you're supposed to critique. In all of these fields, you can find numerous reporters who are functionally controlled by the people they're supposed to keep watch over. Even on my own beat (especially on my own beat!), science reporting, we're captured. The elaborate system of embargoes and privileged press releases set up by scientific journals and scientific agencies ensures that we report not just *what* they want but how and when they want it. We've unwittingly shifted our allegiance from the public we're supposed to serve to the people we're supposed to investigate.

Capture is a bigger threat than even Stigler first realized. Any profession that depends to some degree on objectivity and whose work affects the fortunes of a group of people with power and money is subject to capture. Science, a field in which objectivity is paramount, is far from immune. There's evidence that medical researchers who take money from industry tend see the natural world in a more positive light: In their experiments, drugs seem to work better, patients seem to survive longer, and side effects seem less dangerous. Yet few scientists, even those taking tens or hundreds of thousands of dollars from drug companies or medical-device manufacturers, think they serve any master but Truth with a capital T. That's what worries me the most about capture: You never know when you're a captive.

THE TRIUMPH OF THE VIRTUAL

MIHALY CSIKSZENTMIHALYI
Distinguished Professor of Psychology & Management; founder & codirector, Quality of Life Research Center, Claremont Graduate University; author, A Life Worth Living: Contributions to Positive Psychology

I tried to rank my fears in order of their severity, but soon I realized I would not complete this initial task before the submission deadline, so I decided to use a random number generator to choose among the fears. It turned out not to be a bad choice: basically, the fear that in one or two generations children will grow up to be adults unable to tell reality from imagination. Of course, humanity has always had a precarious hold on reality, but it looks like we're headed for a quantum leap into an abyss of insubstantiality.

I don't know if you have been following the launch of the new 3-D version of one of the major multiplayer video games, replete with monsters, orcs, slavering beasts, and all sorts of unsavory characters brandishing lethal weapons. To survive in this milieu, the player needs fast responses and a quick trigger finger. And now let's reflect on the results of what happens when children start playing such games before entering school and continue to do so into their teens. A child learns about reality through experiences first, not through lectures and books. The incessant warfare he takes part in is not virtual to the child—it is his reality. The events on the screen are more real than the War of Independence or World War II. At a superficial cognitive level they're aware the game is only a virtual reality, but at a deeper, emotional level they know it is not. After all, it is *happening* to them.

It's true that some of the oldest and most popular games are based on forms of mayhem. Chess, for instance, consists in eliminating and immobilizing the enemy forces: infantry, cavalry, messengers, troops mounted on elephants, and the redoubtable queen. (The last was actually a misunderstanding: The Persian inventors of the game gave the most important warrior the title of "Vizier," after the designation for the commanders of the Persian army; the French Crusaders who learned the game as they wandered through the Near East thought the piece was called *Vierge*, after the Virgin Mary; upon their return to Europe, the Virgin became the Queen.) But chess, although it can be an obsession, can never be confused with the rest of reality by a sane person. The problem with the new gaming technology is that it has become so realistic that with enough time and with little competition from the child's environment (which tends to be safe, boring, and predictable), it can erase the distinction between virtual and real. It is then a short step for a young man on the brink of sanity to get ahold of one of the various attack weapons so conveniently available and go on a shooting spree that is a mere continuation of what he has been doing "virtually" for years.

A few decades ago, I started doing research and writing about the impact of indiscriminate television watching, especially on children. Then, as the interactive video games began to appear on the market, it seemed that finally the electronic technology was becoming more child-friendly: Instead of watching passively inane content, children would now have a chance to become engaged in stimulating activities. What I did not have the sense to imagine was that the engagement offered by the new technology would become a Pandora's box containing bait for the reptilian brain to feast on. What scares me now is that children experiencing such reality are going to create a really real world like the one Hieronymus Bosch envisioned—full of spidery creatures, melting objects, and bestial humans.

THE PATIENCE DEFICIT

NICHOLAS G. CARR

Author, The Shallows: What the Internet Is Doing to Our Brains

I'm concerned about time—the way we're warping it and it's warping us. Human beings, like other animals, seem to have remarkably accurate internal clocks. Take away our wristwatches and our cell phones and we can still make pretty good estimates about time intervals. But that faculty can also be easily distorted. Our perception of time is subjective; it changes with our circumstances and our experiences. When things are happening quickly all around us, delays that would otherwise seem brief begin to seem interminable. Seconds stretch out. Minutes go on forever. "Our sense of time," observed William James in his 1890 masterwork *The Principles of Psychology*, "seems subject to the law of contrast."

In a 2009 article in the *Philosophical Transactions of the Royal Society*, the French psychologists Sylvie Droit-Volet and Sandrine Gil described what they call the paradox of time: "Although humans are able to accurately estimate time as if they possess a specific mechanism that allows them to measure time," they wrote, "their representations of time are easily distorted by the context." They describe how our sense of time changes with our emotional state. When we're agitated or anxious, for example, time seems to crawl; we lose patience. Our social milieu, too, influences the way we experience time. Studies suggest, write Droit-Volet and Gil, "that individuals match their time with that of others." The "activity rhythm" of those around us alters our own perception of the passing of time.

Given what we know about the variability of our time sense, it seems clear that information and communication technologies would have a particularly strong effect on personal time perception. After all, they often determine the pace of the events we experience, the speed with which we're presented with new information and stimuli, and even the rhythm of our social interactions. That's long been true, but the influence must be particularly strong now that we carry powerful and extraordinarily fast computers around with us. Our gadgets train us to expect near instantaneous responses to our actions, and we quickly get frustrated and annoyed at even brief delays.

I know that my own perception of time has been changed by technology. If I go from using a fast computer or Web connection to using even a slightly slower one, processes that take just a second or two longer—waking the machine from sleep, launching an application, opening a Web page—seem almost intolerably slow. Never before have I been so aware of, and annoyed by, the passage of mere seconds.

Research on Web users shows that this is a general phenomenon. Back in 2006, a famous study of online retailing found that a large percentage of online shoppers would abandon a retailing site if its pages took four seconds or longer to load. In the years since then, the so-called Four Second Rule has been repealed and replaced by the Quarter of a Second Rule. Studies by companies like Google and Microsoft now find it takes a delay of just 250 milliseconds in page-loading for people to start abandoning a site. "Two hundred fifty milliseconds, either slower or faster, is close to the magic number now for competitive advantage on the Web," a top Microsoft engineer said in 2012. To put that into perspective, it takes about the same amount of time for you to blink an eye.

A recent study of online video viewing provides more evidence of how advances in media and networking technology reduce our patience. Shunmuga Krishnan and Ramesh Sitaraman studied a huge database that documented 23 million video views by nearly 7 million people. They found that people start abandoning a video in droves after a 2-second delay. That won't surprise anyone who has had to wait for a video to begin after clicking the Start button. More interesting is the study's finding of a causal link between higher connection speeds and higher abandonment rates. Every time a network gets quicker, we become antsier. As we experience faster flows of information online, we become, in other words, less patient people. But it's not just a network effect. The phenomenon is amplified by the constant buzz of Facebook, Twitter, texting, and social networking in general. Society's "activity rhythm" has never been so harried. Impatience is a contagion spread from gadget to gadget.

All of this has obvious importance to anyone involved in online media or in running data centers. But it also has implications for how all of us think, socialize, and in general live. If we assume that networks will continue to get faster—a pretty safe bet—then we can also conclude that we'll become more and more impatient, more and more intolerant of even microseconds of delay between action and response. As a result, we'll be less likely to experience anything that requires us to wait, that doesn't provide us with instant gratification. That has cultural as well as personal consequences. The greatest of human works—in art, science, politics—tend to take time and patience both to create and to appreciate. The deepest experiences can't be measured in fractions of seconds.

It's not clear whether a technology-induced loss of patience persists even when we're not using the technology. But I would hypothesize (based on what I see in myself and others) that our sense of time is indeed changing in a lasting way. Digital technologies are training us to be more conscious of and more antagonistic toward delays of all sorts—and perhaps more intolerant of moments of time that pass without the arrival of new stimuli. Because our experience of time is so important to our experience of life, it strikes me that these kinds of technology-induced changes in our perceptions can have broad consequences. In any event, it seems like something worth worrying about, if you can spare the time.

THE TEENAGE BRAIN

SARAH-JAYNE BLAKEMORE
Royal Society University Research Fellow and professor of cognitive neuroscience, University College London; coauthor (with Uta Frith), The Learning Brain: Lessons for Education

Until about fifteen years ago, it was widely assumed that the majority of brain development occurs in the first few years of life. But recent research on the human brain has demonstrated that many brain regions undergo protracted development throughout adolescence and beyond. This advance in knowledge has intensified old worries and given rise to new ones. It is hugely worrying that so many teenagers around the world don't have access to education while their brains are still developing and being shaped by the environment. We should also worry about our lack of understanding of how our rapidly changing world is shaping the developing teenage brain.

Decades of research on early neurodevelopment demonstrated that the environment influences brain development. During the first few months or years of life, an animal must be exposed to particular visual or auditory stimuli for the associated brain cells and connections to develop; in this way, neuronal circuitry is sculpted. This research has focused mostly on early development of sensory brain regions. What about later development of higher-level brain regions, such as prefrontal cortex and parietal cortex, which are involved in decision making, inhibitory control, and planning, as well as social understanding and self-awareness? We know that these brain regions continue to develop throughout adolescence; however, we have very little knowledge about how environmental factors influence the developing teenage brain. This is something that should concern us.

There is some recent evidence from the Dunedin longitudinal study that adolescence represents a period of brain development particularly sensitive to environmental input. This study reported that persistent cannabis use in adolescence had long-lasting negative consequences on a broad spectrum of cognitive abilities in adulthood. This was *not* the case if cannabis use started after the age of eighteen. Could the same be true for other environmental factors—alcohol, tobacco, drug use, diet, medication, Internet usage, gaming? These are all likely to affect the developing brain. The question is how, and we simply don't know the answer.

There's a lot of concern about the hours some teenagers spend online and playing video games. But maybe all this worry is misplaced. After all, throughout history, humans have worried about the effects of new technologies on the minds of the next generation. When the printing press was invented, there was anxiety that reading might corrupt young minds, and the same worries were repeated for the invention of radio and television. Maybe we shouldn't be worried at all. It's possible that the developing brains of today's teenagers will be the most adaptable, creative, multitasking brains that have ever existed. There is evidence—from adults—that playing video games improves a range of cognitive functions, such as divided attention and working memory as well as visual acuity. Much less is known about how gaming, social networking, and so on influence the developing adolescent brain. We don't know whether the effects of new technologies on the developing brain are positive, negative, or neutral. We need to find out.

Adolescence is a period of life when the brain is malleable, and it represents a good opportunity for learning and social development. However, according to UNICEF, 40 percent of the world's teenagers have no access to secondary-school education. The percentage of teenage girls who lack this access is much higher, yet there is strong evidence that the education of girls in developing countries has many significant benefits for family health, population growth rates, child mortality rates, and HIV rates, as well as for women's self-esteem and quality of life. Adolescence represents a time of brain development when teaching and training should be particularly beneficial. I worry about the lost opportunity of denying the world's teenagers access to education.

WHO'S AFRAID OF THE BIG BAD WORDS?

BENJAMIN BERGEN
Associate professor, Department of Cognitive Science, UC San Diego; author, Louder than Words: The New Science of How the Mind Makes Meaning

At around 2:00 P.M. on Tuesday, October 30, 1973, a New York radio station played a monolog by the comedian George Carlin, enumerating and exemplifying in rich detail the seven words ostensibly not allowed on the public airwaves. Soon afterward, the FCC placed sanctions on the radio station for the broadcast, which it deemed "indecent" and "patently offensive." Five years later, the U.S. Supreme Court upheld its decision. In other words, the highest court in the land judged certain words to be so dangerous that even the constitutional right to free speech had to be set aside. But why?

The children, of course. It was to protect the children. According to the Supreme Court, the problem with Carlin's routine was that the obscene words, words describing sexual acts and excretory functions, "may have a deeper and more lasting negative effect on a child than on an adult."

Many of us are afraid of exposing children to taboo language based on the same notion—that somehow certain words can damage young minds. And the well-being of children, were it indeed on the line, would most certainly be a justifiable reason to limit freedom of speech. But the problem is that the Court's premise—that children can be harmed by selected taboo words—does not survive the test of empirical scrutiny. In fact, there are no words so terrible, so gruesomely obscene, that merely hearing them or speaking them poses any danger to young ears.

Taboo words carry no intrinsic threat of harm. Simply referring to body parts or actions involving them do not harm a child. Indeed, the things that taboo words refer to can be equally well identified using words deemed appropriate for medical settings or use around children. And there's nothing about the sound of the words themselves that causes insult to the child's auditory system. Near phonological neighbors to taboo words, words like "fit" and "shuck," do not contaminate the cochlea.

Indeed, which particular words are selected as forbidden is an arbitrary accident of history. Words that once would have earned the utterer a mouthful of soap—expressions like "Zounds!" or "That sucks!" —hardly lead the modern maven to bat an ear. And conversely, words that today rank among the most obscene were at one time commonly used to refer to the most mundane things, like roosters and female dogs. No, the only risk children run by hearing the four-letter words prohibited over the public airwaves is the small chance of broadening their vocabularies. And even this possibility is remote, as anyone can attest who has recently overheard the goings-on in an elementary school playground.

So when the Motion Picture Association of America forbids children from watching the *South Park* movie; when parents instruct children to put their hands over their ears in "earmuff" position; and, indeed, when the FCC levies fines on broadcasters, they aren't protecting children. But they *are* having an effect. Paradoxically, it's these actions we take to shield children from words, with censorship foremost among them, that give specific words their power. And this makes perhaps the best argument that w shouldn't be afraid of exposing children to taboo words. Doin is the best way to take away any perceived threat they pose.

THE CONTEST BETWEEN ENGINEERS AND DRUIDS

PAUL SAFFO
Technology forecaster; consulting professor, School of Engineering, Stanford University

There are two kinds of fools: one who says this is old and therefore good, and the other who says this is new and therefore better. The argument between the two is as old as humanity itself, but technology's exponential advance has made the divide deeper and more contentious than ever. My greatest fear is that this divide will frustrate the sensible application of technological innovation meant to solve humankind's greatest challenges.

The two camps forming this divide need a name, and Druids and Engineers will do. Druids argue that we must slow down and reverse the damage and disruption wrought by two centuries of industrialization. Engineers advocate the opposite: We can overcome our current problems only with the heroic application of technological innovation. Druids argue for a return to the past; Engineers urge us to flee into the future.

The Druid/Engineer divide can be seen in virtually every domain touched by technology. Druids urge a ban on GMOs (genetically modified organisms); Engineers impatiently argue for the creation of synthetic organisms. Environmental Druids seek what the late David Brower called "Earth National Park," while Engineers would take a page from Douglas Adams' lanet-designing Magratheans in *The Hitchhiker's Guide to the laxy*, making a better Earth by fixing all the broken bits. shumanists and Singularitans are Engineers; the Animal

Liberation Front and Ted Kaczynski are Druids. In politics, Libertarians are Engineers, whereas the Greens are Druids. Among religions, Christian fundamentalists are Druids and Scientologists are Engineers.

The gulf between Druid and Engineer makes the divide between C. P. Snow's Two Cultures seem like a crack in the sidewalk. The two camps do not merely hold different worldviews; they barely speak the same language. A recent attempt to sequester oceanic carbon by dumping iron dust in the Pacific off British Columbia intrigued Engineers but alarmed Druids, who considered it an act of intentional pollution. Faced with uncertainty or crisis, Engineers instinctively hit the gas; Druids hit the brake.

The pervasiveness of the Druid/Engineer divide and the stubborn passions demonstrated by both sides reminds me of that old warrior-poet Archilochus and his hedgehog/fox distinction, revived and elaborated upon by Isaiah Berlin. Experience conditions us toward being Engineers or Druids just as it turns us into hedgehogs or foxes. Engineers tend to be technologists steeped in physics and engineering. Druids are informed by anthropology, biology, and the Earth sciences. Engineers are optimists: Anything can be fixed, given enough brain power, effort, and money. Druids are pessimists: No matter how grand the construct, everything eventually rusts, decays, and erodes to dust.

Perhaps the inclination is even deeper. Some years back, the five-year-old daughter of a venture capitalist friend announced, upon encountering an unfamiliar entree at the family table, "It's new, and I don't like it." A Druid in the making, that became her motto all through primary school—and for all I know, it still is today.

We live in a time when the loneliest place in any debate is the middle, and the argument over technology's role in our future is no exception. The relentless onslaught of novelties technologic and otherwise is tilting individuals and institutions alike toward becoming Engineers or Druids. It is a pressure we must resist, for to be either a Druid or an Engineer is to be a fool. Druids can't revive the past, and Engineers cannot build technologies that do not carry hidden trouble.

The solution is to claw our way back to the middle, and a good place to start is by noting one's own Druid/Engineer inclinations. Unexamined inclinations amount to dangerous bias, but once known, the same inclination can become the basis for powerful intuition. What is your instinctive reaction to something new; is anticipation or rejection your default? Consider autonomous highway vehicles: Druids fear that robot cars are unsafe; Engineers wonder why humans are allowed to drive at all.

My worry is that collective minds change at a snail's pace, whereas technology races along an exponential curve. I fear we will not rediscover the middle ground in time to save us from our myriad folly. My inner Engineer is certain that a new planetary meme will arrive and bring everyone to their senses, but my gloomy Druid tells me that we will be lucky to muddle our way through without killing ourselves off or ushering in another Dark Age. I'll be happy if both are a little right—and a little wrong.

"SMART"

EVGENY MOROZOV
Contributing editor, The New Republic; *syndicated columnist; author*, To Save Everything, Click Here: The Folly of Technological Solutionism

I worry that as the problem-solving power of our technologies increases, our ability to distinguish between important and trivial or even nonexistent problems diminishes. Just because we have "smart" solutions to fix every single problem under the sun doesn't mean that all of those problems deserve our attention. In fact, some of them may not be problems at all; that certain social and individual situations are awkward, imperfect, noisy, opaque, or risky might be by design. Or, as the geeks like to say, some bugs are not bugs, some bugs are features.

I find myself preoccupied with the invisible costs of "smart" solutions in part because Silicon Valley mavericks are not lying to us: Technologies are becoming not only more powerful but also more ubiquitous. We used to think that, somehow, digital technologies lived in a national reserve of some kind; first we called this imaginary place "cyberspace" and then we switched to the more neutral label of "Internet." It's only in the last few years, with the proliferation of geolocational services, self-driving cars, and smart glasses, that we grasped that such national reserves were perhaps a myth and that digital technologies would be everywhere: in our fridges, on our belts, in our books, in our trash bins.

All this smart awesomeness will make our environment more plastic and more programmable. It will also tempt us to design out all imperfections—just because we can!—from our interactions, social institutions, politics. Why have an expensive law enforcement system, if we can design smart environments where no crimes are committed simply because those people deemed "risky"—based, no doubt, on their online profiles—are barred from access and thus unable to commit crimes in the first place? So we are faced with a dilemma: Do we want some crime or no crime? What would we lose—as a democracy—in a world without crime? Would our debate suffer, as the media and courts would no longer review the legal cases? This is an important question that I'm afraid Silicon Valley, with its penchant for efficiency and optimization, might not get right.

Or take another example: If, through the right combination of reminders, nudges, and virtual badges, we can get people to be "perfect citizens"—recycle, show up at elections, care about urban infrastructure—should we take advantage of the possibilities offered by smart technologies? Or should we, perhaps, accept that slacking off and idleness, in small doses, are productive in that they create spaces and openings where citizens can still be appealed to by deliberation and moral argument, not just the promise of a better shopping discount courtesy of their smartphone app?

If problem solvers can get you to recycle via a game, would they even bother with the less effective path of engaging you in moral reasoning? The difference is that those people earning points in a game might end up not knowing anything about the "problem" they were solving, while those who had been through the argument would have a tiny chance of grasping the issue's complexity and doing something that would matter in the years to come, not just today.

Alas, smart solutions don't translate into smart problem solvers. In fact, the opposite might be true: Blinded by the awesomeness of our tools, we might forget that some problems and imperfections are just the normal costs of accepting the social contract of living with other human beings, treating them with dignity, and ensuring that, in our recent pursuit of a perfect society, we do not shut the door to change. Change usually happens in rambunctious, chaotic, and imperfectly designed environments; sterile environments, where everyone is content, are not known for innovation, of either the technological or the social variety. When it comes to smart technologies, there's such a thing as too "smart," and it isn't pretty.

THE STIFLING OF TECHNOLOGICAL PROGRESS

DAVID PIZARRO
Associate professor of psychology, Cornell University

It is increasingly clear that human intuitions—particularly our social and moral intuitions—are ill equipped to deal with the rapid pace of technological innovation. We should be worried that this will hamper the adoption of technologies that might otherwise be of practical benefit to individuals and great benefit to society. Here's an example: My e-mail provider has long been able to generate targeted advertisements based on the content of my e-mail. But it can now also suggest a calendar entry for an upcoming appointment mentioned in an e-mail, track my location as the appointment approaches, alert me about when to leave, and initiate driving directions to get me there on time.

It feels natural to say that Google "reads my e-mail" and that it "knows where I have to be." We can't help but interpret this automated information through the lens of our social intuitions, and we end up perceiving agency and intentionality where there is none. So even if we know that no human eyes have seen our e-mails, it can still feel, well, creepy—as if we're not quite sure that there isn't someone going through our stuff, following us around, and possibly talking about us behind our back. Unsurprisingly, many view these services as a violation of privacy, even when there's no agent doing the "violating." The adoption of such technologies has suffered for these reasons.

These social intuitions interfere with the adoption of technologies offering more than mere convenience. For instance, the technology for self-driving cars exists now and promises that thousands of lives may be saved each year because of reduced traffic collisions. But the technology depends fundamentally on the ability to track one's precise location at all times. This is just creepy enough that a lot of people will likely avoid the technology and opt for the riskier option of driving themselves.

Of course, we're not necessarily at the whims of our psychological intuitions. Given enough time we can (and do) learn to set them aside when necessary. However, I doubt we can do so quickly enough to match the current speed of technological innovation.

THE RISE OF ANTI-INTELLECTUALISM AND THE END OF PROGRESS

TIM O'REILLY
Founder and CEO of O'Reilly Media

For many in the techno-elite, even those who don't entirely subscribe to the unlimited optimism of the Singularity, the notion of perpetual progress and economic growth is somehow taken for granted. As a former classicist turned technologist, I've lived with the shadow of the fall of Rome, the failure of its intellectual culture, and the stasis that gripped the Western world for the better part of 1,000 years. What I fear most is that we will lack the will and foresight to face the world's problems squarely and will instead retreat from them into superstition and ignorance.

Consider how in A.D. 375, after a dream in which he was whipped for being a "Ciceronian" rather than a Christian, St. Jerome resolved to abandon the classical authors and restrict himself to Christian texts, and how in A.D. 415 the Christians of Alexandria murdered the philosopher and mathematician Hypatia—and realize that, at least in part, the Dark Ages were not something imposed from without, a breakdown of civilization due to barbarian invasions, but a choice, a turning away from knowledge and discovery into a kind of religious fundamentalism. Now consider how conservative elements in American religion and politics refuse to accept scientific knowledge and deride their opponents for being "reality based," and ask yourself, "Could that ideology come to rule the most powerful nation on Earth? And if it did, what would be the consequences to the world?"

History teaches us that conservative, backward-looking movements often arise under conditions of economic stress. As the world faces problems ranging from climate change to the demographic cliff of aging populations, it's wise to imagine widely divergent futures. Yes, we may find technological solutions that propel us into a new Golden Age of robots, collective intelligence, and an economy built around "the creative class." But it's at least as probable that as we fail to find those solutions quickly enough, the world falls into apathy, disbelief in science and progress, and, after a melancholy decline, a new Dark Age.

Civilizations do fail. We have never seen one that hasn't. The difference is that the torch of progress has, in the past, always passed to another region of the world. But we now for the first time have a single, global civilization. If it fails, we all fail together.

ARMAGEDDON

TIMOTHY TAYLOR
Archaeologist, professor of the prehistory of humanity, University of Vienna; author, The Artificial Ape: How Technology Changed the Course of Human Evolution

As I get older, I no longer slam the door on Jehovah's Witnesses or Mormons. Nor do I point to the large icon of the winged St. John of Patmos (putative author of the Book of Revelation) at the head of the stairs and claim falsely to be Bulgarian Orthodox—I find it hard to keep a straight face. The last pair of young Midwest Mormons, clinging to each other in the alien country of Europe, had clearly heard of neither Bulgaria nor Orthodoxy.

My change of heart crystallized when two Jehovah's Witnesses knocked and asked whether I was optimistic. A genuine smile crossed my face as I affirmed that I was. The one in charge (the other being the apprentice, as is usual) whipped out a copy of *The Watchtower*, the front cover of which showed a spectacular mushroom cloud emblazoned with the headline "ARMAGEDDON!" He suggested that my casual optimism might be misplaced.

I began to see his point, though not from his point of view. Armageddon for Jehovah's Witnesses immediately precedes the Last Judgment and is the prophesied final battle against the Antichrist. It probably takes its name from the ancient city of Megiddo, Israel—an Old Testament byword for big, locally apocalyptic battles. Thutmose III of Egypt put down a Canaanite rebellion there in 1457 B.C., and in 609 B.C. the Egyptians again triumphed, defeating the Iron Age Judeans. Megiddo is a doubly strategic location: Locally, it controls the route from the eastern Mediterranean coast at Mount Carmel through to the Jordan Valley; regionally, it occupies a relatively narrow tract of habitable land that bridges Africa and Eurasia.

Neolithic farmers first settled Megiddo itself around 7000 B.C., but the area was occupied far earlier. Excavations at the rock shelter at Qafzeh have produced stone tools, pigment-stained shell ornaments, and the ceremonial burials of over a dozen people, dating to 92,000 years ago. The nearby cave of Tabun contains tools from *Homo erectus* that date to half a million years ago and, from around 200,000 years ago onward, the remains of Neanderthals, a different human species who eventually came into competition with groups of anatomically modern *Homo sapiens* filtering northward out of Africa at around the time of the Qafzeh burials.

These facts about our evolutionary and cultural prehistory, established by international transdisciplinary scientific projects, are utterly rejected by fundamentalists. Archbishop Ussher of Armagh claimed that Adam and Eve, the first people, were made directly by God on the sixth day of Creation, late October 4004 B.C. (only to be driven from Paradise on Monday, the 10th of November of the same year). By the Witnesses' calculations, the world is more than a whole century older, with Adam created in 4129 B.C. But neither understanding leaves time for the early farming phase of Megiddo. Or for biological evolution. Or plate tectonics. Or anything, really, that has been established empirically by the historical sciences.

If this were simply a disagreement about the scale of the past or the processes that operated in it, it might be little more than a frustration for archaeologists, paleontologists, and geologists. But it seems to me that belief in a fixed and recent start to the world is invariably matched by belief in an abrupt and (usually) imminent end. The denial of antiquity, and of Darwinian evolution, psychologically defines the form and scope of any imagined future. And that has implications for the way individuals and communities make decisions about resource management, biodiversity, population control, and the development of technologies.

None of those things matter to Armageddonists. By "Armageddonist," I mean an adherent—irrespective of fine-grade distinctions in the eschatologies, or "end-times thinking," of particular religious faiths—of the idea that (1) everything will be miraculously solved at some definite point in the (near) future; (2) the solution will create eternal winners and losers; and (3) the only real concern in life is which side we find ourselves on when the battle (however concrete or abstract) commences. A coming Battle of Armageddon is anticipated by Jehovah's Witnesses, the Islamic Ahmadiyya community, the Seventh-day Adventists, Christian dispensationalists, Christadelphians, and members of the Baháì faith. More broadly, the idea of apocalyptic renewal at "the end of days" is shared by the Church of Jesus Christ of Latter-day Saints, Islam, Judaism, and most other religions (with the noteworthy exception of Zen Buddhism).

Obviously, moral distinctions should be made between religionists who blow people up and stone apostates to death and those who do not. There is also a distinction between faiths that reject science wholesale and those that do not. The Catholic Church has long eschewed burning heretics in the *auto-da-fé*, and it never rejected Darwin's view of human evolution: Already in 1868, John Henry—later Cardinal—Newman found it potentially compatible; positive acceptance has emerged gradually since 1950. But Catholicism retains a firm belief in a precisely delimited future leading to the Last Judgment. Rome's opposition to contraception connects to this; the globe does not need to be sustainable in perpetuity if all will be renewed.

Admittedly, scientists can be as good as priests at scaring people into despairing visions of the end of the world. If they do, they should not be surprised when people are attracted by options that include salvation. But faced with claims that run counter to empirical evidence, those of us of an Enlightenment disposition can never afford to appease, relativize, or bow to "religious sensitivities," whether through misplaced tact or (more honorably, given the penalties now on display in some jurisdictions) a sense of self-preservation. The science deniers will show no reciprocal respect to reason.

We should be worried about Armageddon not as a prelude to an imaginary divine Day of Judgment but as a particular, maladaptive mindset that seems to be flourishing despite unparalleled access to scientific knowledge. Paradoxically, it may flourish *because* of this. Ignorance is easy and science is demanding, but, more tellingly, being neither tribal nor dogmatic, science directly challenges ideologs who need their followers to believe them infallible. We should not underestimate the glamour and influence of anti-science ideologies. Left unchecked, they could usher in a new intellectual Dark Age.

This is what has happened in Nigeria, whose populous north now languishes under the baleful Islamist ideology of *Jamaatu Ahlus-Sunnah Lidda'Awati Wal Jihad*, known more pithily as Boko Haram (loosely, "Books are Forbidden"). Boko Haram maintains that we inhabit a flat, 6,000-year-old Earth and that the disk-shaped sun, which is smaller, passes over it daily. Heresy (such as belief in evolution, or that rain is predicated on evaporation) can lead to a judgment of apostasy, which is punished by death.

Reason having fled, the people are increasingly struggling with malnutrition, drought, and disease. But that, too, feeds back into the ideology: In their death notes, the bomb-murderers of Boko Haram articulate detailed visions of the rewards of an afterlife that must seem tragically attractive.

These days when I stand at the doorstep with a winged apocalyptic evangelist behind me and an anoraked one before me, I feel obliged to put over a scientific view of the world, civilly, firmly, clearly. Even with optimism. After all, I am not so in thrall to Armageddon as to think that a coming intellectual Dark Age, even a global one, would last forever.

SUPERSTITION

MATT RIDLEY
Science writer; founding chairman, International Centre for Life; author, The Rational Optimist

I worry about superstition. Rational optimists spend much of their energy debunking the charlatans who peddle false reasons to worry. So what worries me most are the people who make others worry about the wrong things, the people who harness the human capacity for superstition and panic to scare us into doing stupid things: banning genetically modified foods, teaching children that the Earth is 6,000 years old, preventing the education of girls, erecting barriers against immigrants or free trade, preventing fossil fuels taking the pressure off rain forests—that sort of thing.

Superstition can help bring down whole civilizations. As Rome collapsed, or in the Ming Empire, or under the Abbasid Caliphate, the triumph of faith over reason played a large part in turning relative into absolute decline in living standards. By faith, I mean argument from authority.

There is a particular reason to worry that superstition is on the rise today—a demographic reason. As Eric Kaufmann documents in his book *Shall the Religious Inherit the Earth?*, the fundamentalists are breeding at a faster rate than the moderates within all the main religious sects: Sunni, Shia, Jewish, Catholic, Protestant, Mormon, Amish. The differential is great and growing.

Fortunately, children do not always do what their parents tell them. Millions of indoctrinated fundamentalist children will rebel against their faith and embrace reason and liberty, especially in the age of the Internet, mobile telephony, and social networks. But there's another demographic trend: the declining global birthrate, with the number of children per woman plummeting to two or less in more and more countries. In a world with generally high birthrates, a breeding frenzy among fundamentalists would not matter much, but in a world with low birthrates the effect could be startling. If secular folk don't breed while superstitious ones do, the latter will soon dominate.

It's not just religious superstition that bothers me. Scientific superstition seems to be on the rise, too, though not because of demography. Science as an institution, as opposed to a philosophy, has long had a tendency to drift toward faith—argument from authority—as well. Consider the example of eugenics in the first half of the last century, or Lysenkoism in the Soviet Union, or Freudianism, or the obsession with dietary fat that brooked no dissent in the 1970s and 1980s. Dissidents and moderates are all too often crowded out by fundamentalists when science gets political. Fortunately, science has a self-correcting tendency, because it is dispersed among different and competing centers. Dread the day when science becomes centralized.

RATS IN A SPHERICAL TRAP

GREGORY BENFORD

Professor of physics & astronomy, emeritus, UC Irvine; science fiction author, Anomalies

One iconic image expresses our existential condition: the pale blue dot. That photograph of Earth the *Voyager 1* spacecraft took in 1990 from 6 billion kilometers away told us how small we are. What worries me is that that dot may be all we ever have, all we can command, for the indefinite future. Humanity could become like rats stuck on the skin of our spherical world, which would look more and more like a trap.

Imagine: We've had our burgeoning history here and used up many resources . . . so what happens when they run out? Valuable things like metals, rare earths, fertilizers, and the like are already running low. More things will join the list.

Voyager has flown outward for thirty-six years, a huge return on the taxpayers' investment. It is the first probe that will leave the solar system and the farthest man-made object from Earth. Voyager is now exploring the boundary between our little system and the vast interstellar space beyond.

It can instruct us, still, about our more pressing problems, as Carl Sagan pointed out: "Think of the rivers of blood spilled by all those generals and emperors so that in glory and triumph they could become the momentary masters of a fraction of a dot." They fought over resources we could exhaust within the next century or two. *Voyager*'s perspective also suggests an answer: There's a whole solar system out there. Sagan pondered that aspect, too: "There is nowhere else, at least in the near future, to which our species could migrate. Visit, yes. Settle, not yet." That is still true, but the vast solar system can help us. I worry that we will miss this opportunity.

After all, great civilizations have turned away from horizons before. In the 1400s, China stopped trading and exploring in Africa and Arabia out of concern that this introduced unsettling elements into their society. Their nine-masted ships were 100 meters long and carried animals like giraffes back to China to amuse the emperor. China could have found America and ruled the oceans, but it chose to stay home.

We may be entering a similar age. This century will doubtless see our population rise from its current 7 billion souls to 9 or 10 billion. Climate change will wrack economies and nations. The bulk of humanity has large economic ambitions that will strain our world to satisfy. With the United States imitating Europe in its evolution into an entitlement state, it will have less energy to maintain world order. Amid constant demands for more metals, energy, food, and all the rest, it seems clear that we can expect conflicts among those who would become "momentary masters of a fraction of a dot."

But we see in our skies the resources that can aid the bulk of humanity. With entrepreneurs now pulsing with energy, we have plausible horizons and solutions visible. SpaceX (Space Exploration Technologies Corporation), founded by former PayPal entrepreneur Elon Musk, now delivers cargo to the International Space Station. SpaceX became the first private company to successfully launch and return a spacecraft from orbit, on December 8, 2010, and Musk remarked on his larger agenda, the economic opening of space, "We need to figure out how to have the things we love, and not destroy the world." The black expanse over our heads promise places where our industries can use resource extraction, zero-gravity manufacturing, better communications, perhaps even energy harvested in great solar farms and sent down to Earth. Companies are already planning to do so: Bigelow Aerospace (orbital hotels), Virgin Galactic (low Earth orbit tourism), Orbital Technologies (a commercial manufacturing space station), and Planetary Resources, whose goal is to develop a robotic asteroid mining industry.

Barely visible now is an agenda we can carry out in this century to avoid calamity, those rivers of blood and anguished need. We Americans especially know from history how to open new territory. Historically, coal and the railroads enabled much of the Industrial Revolution. Both came from the underlying innovation of steam engines. Coal was the new wonder fuel, far better than wood though harder to extract, and it made continental-scale economies possible. Synergistically, coal drove trains that in turn carried crops, crowds, and much else.

A similar synergy may operate to open the coming interplanetary economy, this time wedding nuclear rockets and robotics. These could operate together, robot teams carried by nuclear rockets to far places—and usually without humans, who would compromise efficiency. Mining and transport have expanded the raw materials available to humanity, and the rocket/robot synergy could do so again. As such fundamentals develop in space, other businesses can arise on this base, including robotic satellite repair/maintenance in high orbits, mining of helium-3 on the moon, and metal mining of asteroids. Finally, perhaps snagging comets for volatiles in the outer solar system will enable human habitats to emerge in hollowed-out asteroids and on Mars and beyond.

Nothing has slowed space development more than the high

price of moving mass around the solar system. Using two stages to get into low Earth orbit may make substantial improvements, and beyond that the right answer may lie in nuclear rockets. These have been developed since the 1960s and could be improved still further. Lofting them into orbit "cold"—that is, before turning on the nuclear portion—may well erase the environmental issues. Fuel fluids can be flown up separately, for attachment to the actual rocket drive. Then the nuclear segment can heat the fuel to very high temperatures. Economically this seems the most promising way to develop an interplanetary economy for the benefit of humanity.

Such ideas have been tried out in the imaginative lab of science fiction, exploring how new technologies could work in a future human context. Kim Stanley Robinson's 2012 visionary novel, *2312*, portrays such a solar system economy. Our reborn entrepreneurial space culture emerged from an earlier epoch that saw space as the inevitable next horizon. The fresh industrial reality is already making a larger cultural echo in novels, films, and much else. It's a space era again.

Sagan spoke often of how the view from space gave us perspective on our place in the cosmos. That started with Apollo 8's 1968 swing around our moon and its backward look at Earth. Many felt, looking at those photos, that future exploration of space should focus on ways to protect Earth and extend human habitation beyond it. Sagan had the idea of turning *Voyager* to look back at ourselves and told us to take the larger perspective in his *Pale Blue Dot: A Vision of the Human Future in Space.*

That first flowering into space set a tone we should embrace. In the end, history may resemble a zero-sum game ruled by resources. We can win such a game only by breaking out of its assumptions.

Babylonian astrologers thought the stars governed our fates. A thousand years ago, societies were largely religious and prayed to the skies for their salvation. We can seek our futures there now as well. And an age might come when we shall govern the fate of stars.

THE DANGER FROM ALIENS

SETH SHOSTAK
Senior astronomer, SETI Institute; author,
Confessions of an Alien Hunter

The recent reset of the long-count Mayan calendar didn't end the world. But there are serious scientists who worry that Armageddon could soon be headed our way, although from a different quarter—an attack by malevolent extraterrestrial beings.

The concern is that future radio broadcasts to the stars, intended to put us in touch with putative aliens, might carelessly betray our presence to a warlike society and jeopardize the safety of Earth. The physicist Stephen Hawking has weighed in on this dreadful possibility, suggesting that we should be careful about sending signals that could trigger an aggressive reaction from some highly advanced race of extraterrestrials.

It all sounds like shabby science fiction, but even if the probability of disaster is low, the stakes are high. Consequently, some cautious researchers argue that it's best to play safe and keep our broadcasts to ourselves. Indeed, they urge a worldwide policy of restraint and relative quiet. They would forbid the targeting of other star systems with transmissions of greater intensity than the routine radio and television that inevitably leak off our planet.

That sounds like a harmless precaution, and who would quibble about inexpensive insurance against the possible obliteration of our world. But this is one worry I don't share. Even more, I believe the cure is more deadly than the disease. To fret about the danger of transmissions to the sky is both too late and too little. Worse, it will endlessly hamstring our descendants.

Ever since the Second World War, we've been broadcasting high-frequency signals that can easily penetrate Earth's ionosphere and seep into space. Many are television, FM radio, and radar. And despite the fact that the most intense of these transmissions sport power levels of hundreds of thousands of watts or more, they dwindle to feeble static at distances measured in light-years. Detecting them requires a very sensitive receiving setup.

As an example of the difficulty, consider an alien society that wields an antenna comparable to the Arecibo telescope in Puerto Rico—1,000 feet in diameter and the largest single-element radio telescope on Earth. This antenna would be unable to pick up our television broadcasts even from the distance of Proxima Centauri, which, at 4.2 light-years, is our nearest stellar neighbor. And it's improbable that we have cosmic confreres this close, or even ten or twenty times farther. Astronomers such as Frank Drake and the late Carl Sagan have estimated that the nearest Klingons (or whatever their species) are at least a few hundred light-years away. Our leakage signals—when they eventually reach that far—will be orders of magnitude weaker than any our best antennas could detect.

Such arguments might appear to justify the suggestion by the self-appointed defenders of Earth that we need not fear our current broadcasts; they will be undetectably weak. But they claim that we should concern ourselves with deliberate, highly targeted (and therefore highly intense) transmissions. We can continue to enjoy our sitcoms and shopping channels, but we should forbid anyone from shouting in the galactic jungle.

There's a serious flaw in this apparently plausible reasoning. Any society able to do us harm from the depths of space is not at our technological level. We can confidently assume that a culture able to project force to someone else's star system is at least several centuries in advance of us. This statement is independent of whether you believe that such sophisticated beings would be interested in wreaking havoc and destruction. We speak only of capability, not motivation.

Therefore it's reasonable to expect that any such advanced beings, fitted out for interstellar warfare, will have antenna systems far larger than our own. In the second half of the 20th century, the biggest of the antennas constructed by Earthly radio astronomers increased by a factor of 10,000 in collecting area. It hardly strains credulity to assume that Klingons hundreds or thousands of years farther down the technological road will possess equipment fully adequate to pick up our leakage. Consequently, the signals we send willy-nilly into the cosmos—most especially our strongest radars—are hardly guaranteed to be "safe."

There's more. Von R. Eshleman, a Stanford University engineer, pointed out decades ago that by using a star as a gravitational lens you can achieve the ultimate in telescope technology. This idea is a straightforward application of Einstein's general theory of relativity, which predicts that mass will bend space and affect the path of light beams. The prediction is both true and useful: Gravitational lensing has become a favored technique for astronomers who study extremely distant galaxies and dark matter.

However, there's an aspect of this lensing effect that's relevant to interstellar communication: Imagine putting a telescope (radio or optical) onto a rocket and sending it to the sun's gravitational focus—roughly twenty times the distance of Pluto. When aimed back at the sun, the telescope's sensitivity will be increased by thousands or millions of times, depending on wavelength. Such an instrument would be capable of detecting even low-power signals (far weaker than your local top-forty

FM station) from 1,000 light-years away. At the wavelengths of visible light, this setup would be able to find the street lighting of New York or Tokyo from a similar remove. Consequently, it's indisputable that any extraterrestrials with the hardware necessary to engage in interstellar warfare will be able to heft telescopes to the comparatively piddling distance of their home star's gravitational focus.

The conclusion is simple: It's too late to worry about alerting the aliens to our presence. That information is already en route at the speed of light, and alien societies only slightly more accomplished than our own will easily notice it. By the 23rd century, these alerts to our existence will have washed across a million star systems. There's no point in fretting about telling the aliens we're here. The deed has been done, and the letter's in the mail.

But what about a policy to limit our future leakage? What about simply calming the cacophony so we don't continue to blatantly advertise our presence? Maybe our transmissions of the past half-century will somehow sneak by the aliens.

Forget it. Silencing ourselves is both impossible and inadvisable. The prodigious capability of a gravitational-lens telescope means that even the sort of low-power transmissions ubiquitous in our modern society could be detectable. And would you really want to turn off the radar sets down at the airport, or switch off city streetlights? Forever?

In addition, our near-term future will surely include many technological developments that will unavoidably be visible to other societies. Consider powersats—large arrays of solar cells in orbit around the Earth that could provide us with nearly unlimited energy without the noxious emissions or environmental damage. Even in the best cases, such devices would backscatter hundreds or thousands of watts of radio noise into space. Do we want to forbid such beneficial technologies until the end of time?

Yes, some people are worried about being noticed by other galactic inhabitants—beings that might threaten our lifestyle or even our world. But that's a worry without a practical cure, and the precautions that some urge us to take promise more harm than good. I, for one, have let this worry go.

AUGMENTED REALITY

WILLIAM POUNDSTONE
Journalist; author, Are You Smart Enough to Work at Google?

I worry about augmented reality. It's an appealing technology that seems completely inevitable in the next twenty years. You'll wear eyeglasses that are no bulkier than the regular kind, or maybe even contact lenses, and they'll overlay all sorts of useful information on your field of view. You could have an interactive map, a live news ticker, or notifications of messages—anything you can get on a screen now.

The game-changing thing is not the local-specific overlays (which we've already got) but the utter privacy. No one else would know you're checking scores in a business meeting or playing video games in class. These may sound like silly examples, but I don't think they are. How often have you not checked your messages because it wasn't quite socially acceptable to pull out a phone? With eyeglass-mounted augmented reality, all the inhibitions will be gone.

I'm not worried about safety—maybe cars will drive themselves by then. I worry about a world in which everyone is only pretending to pay attention. Our social lives are founded on a premise that has always been too obvious to need articulation: that people attend to the people immediately around them. To not do that was to be rude, absent-minded, or even mentally ill. That's coming to describe us all. We're heading toward a Malthusian catastrophe. Consumer-level bandwidth is still growing exponentially, while our ability to deal with seductive distractions is stable or at best grows arithmetically. We will need to invent a new social infrastructure to deal with that, and I worry that we don't have much time to do it.

TOO MUCH COUPLING

STEVEN STROGATZ

Schumann Professor of Applied Mathematics, Cornell University; author, The Joy of *x*: A Guided Tour of Math, from One to Infinity

In every realm where we exist as a collective—in society, in the global economy, on the Internet—we are blithely increasing the coupling between us, with no idea what that might entail.

"Coupling" refers to the ability of one part of a complex system to influence another. If I put 100 metronomes on the floor and set them ticking, they'll each do their own thing, swinging at their own rhythm. In this condition, they're not yet coupled. Because the floor is rigid, the metronomes can't feel each other's vibrations—at least, not enough to make a difference. But now place them all on a movable platform, like the seat of a child's swing. The metronomes will start to feel each other's jiggling; the swing will start to sway, imperceptibly at first but enough to disturb each metronome and alter its rhythm. Eventually the whole system will synchronize, with all the metronomes ticking in unison. By allowing the metronomes to impose themselves on each other through the vibrations they impart to the movable platform, we have coupled the system and changed its dynamics radically.

In all sorts of complex systems, this is the general trend: Increasing the coupling between the parts seems harmless enough at first. But then, abruptly, when the coupling crosses a critical value everything changes. The exact nature of the altered state isn't easy to foretell. It depends on the system's details. But it's always something qualitatively different from what came before. Sometimes desirable, sometimes deadly.

I worry that we're playing the coupling game with ourselves, collectively. With our cell phones and GPS trackers and social media, with globalization, with the coming Internet of things, we're becoming more tightly connected than ever. Of course, maybe that's good. Greater coupling means faster and easier communication and sharing. We can often do more together than apart.

But the math suggests that increasing coupling is a siren's song. Too much makes a complex system brittle. In economics and business, the wisdom of the crowd works only if the individuals within it are independent, or nearly so. Loosely coupled crowds are the only wise ones.

The human brain is the most exquisitely coupled system we know of, but the coupling between different brain areas has been honed by evolution to allow for the subtleties of attention, memory, perception, and consciousness. Too much coupling produces pathological synchrony, the rhythmic convulsions and loss of consciousness associated with epileptic seizures.

Propagating malware, worldwide pandemics, flash crashes—all symptoms of too much coupling. Unfortunately, it's hard to predict how much coupling is too much. We know only that we want more, and that more is better . . . until it isn't.

HOMOGENIZATION OF THE HUMAN EXPERIENCE

SCOTT ATRAN
Anthropologist, National Center for Scientific Research, Paris; author, Talking to the Enemy: Faith, Brotherhood, and the (Un)Making of Terrorists

More than half a million years ago, the Neanderthal and human branches of evolution began to split from our common ancestor *Homo erectus.* Neanderthal, like *erectus* before, spread out of Africa and across Eurasia. But our ancestors, who acquired fully human structures of brain and body about 200,000 years ago, remained stuck in the savanna grasslands and scrub of eastern, then southern Africa. Recent archaeological and DNA analyses suggest that our species may have tottered on the brink of extinction as recently as 70,000 years ago, dwindling to fewer than 2,000 souls. Then, in the geological blink of an eye, they became us, traipsing about on the moon and billions strong.

How did it happen? No real evidence has emerged for dramatic change in general anatomy of the human body and brain or in basic capacities for physical endurance and perception. The key to this astounding and bewildering development may have been a mutation in the computational efficiency of the brain to combine and process concepts—a recursive language of thought and theory of mind—that led to linguistic communication about possible worlds and to a mushrooming cultural cooperation and creativity within and between groups to better compete against other groups.

From the evanescent beauty of sand paintings by Australian aboriginals and Native Americans to the great ziggurats and pyramids of ancient Mesopotamia, India, and Mesoamerica that functioned mainly to stimulate imagination—and from the foragers, herders, cultivators, warriors, and innovators of New Guinea, the Amazon, Africa, and Europe—a startling multiplicity of social and intellectual forms emerged to govern relations between people and nature.

By the time of Christ, four great neighboring polities spanned Eurasia's middle latitudes along the trading network known as the Silk Road: the Roman Empire; the Parthian Empire, centered in Persia and Mesopotamia; the Kushan Empire of Central Asia and Northern India; and the Han Empire of China and Korea. Along the Silk Road, Eurasia's three universalist moral religions—Judaism, Zoroastrianism, and Hinduism—mutated from their respective territorial and tribal origins into the three proselytizing, globalizing religions vying for the allegiance of our species: Christianity, Islam, and Buddhism. The globalizing religions created two new concepts in human thought: individual free choice and collective humanity. People not born into these religions could, in principle, choose to belong (or remain outside) without regard to ethnicity, tribe, or territory. The mission of these religions was to extend moral salvation to all peoples, not just to a "Chosen People" that would light the way for others.

Secularized by Europe's Enlightenment, the great quasi-religious *isms* of modern history—colonialism, socialism, anarchism, fascism, communism, democratic liberalism—harnessed industry and science to continue on a global scale the human imperative of cooperate to compete, or kill massively to save the mass of humanity.

Soon, from the multiplicity of human cultural forms, there will likely be only one or a few. This is the way of evolution, some scientists say, for in the struggle for life only conquering or isolated forms survive. But the humanists among them hope that with vigilant intellectual and political effort, democracy, reason, human rights, and happiness will flourish in our one interdependent world. Yet science also teaches us that greater interdependence can also lead to greater vulnerability to cascading and catastrophic collapse from single unanticipated events. The two great world wars and the global economic crises of the last century could well be indicators of far greater risks to come.

Calls to "tolerate diversity" and "respect nature" are pale acknowledgments of the problem. Anthropology has virtually disappeared as a science of cultures, with the emphasis now either on personal experience in some slightly exotic place, like contemplating one's own genitals the way others do or don't, or on the cultural dynamics of the mundane, like bench sitters or subway riders. Psychology today considers that comparing, say, American with Chinese college students speaks to the breadth and depth of the human experience, although the apparent differences can usually be erased by a mere change in priming or framing (just suggesting that one might think about something one way rather than another). This tells us little about the true scope and limits of human thought and behavior. And political correctness, which attempts to enforce tolerance, creates bigotry over trivialities.

As the death of languages and cultures proceeds geometrically and further efforts at domestication of nature lead to further alienation from it (including establishment of "nature reserves" and efforts to control climate change), we may find our global culture imploding to a point of precariousness not so different from what it was 70,000 years ago when the human experience began with a bang.

ARE WE HOMOGENIZING THE GLOBAL VIEW OF A NORMAL MIND?

P. MURALI DORAISWAMY
Professor of psychiatry, Duke University Medical Center; member, Duke Institute for Brain Sciences

Should we worry about the consequences of exporting America's view of an unhealthy mind to the rest of the world?

Biologists estimate that there may be between 1.5 million and 5 million subspecies of fungi, though only 5 percent (of the lower estimate) are currently categorized. To outsiders, it may appear that America's classification of mental disorders is not too different. At the turn of the century, psychiatric disorders were mostly categorized into neuroses and psychoses. In 1952, the first version of the Diagnostic and Statistical Manual of Mental Disorders (DSM), the psychiatric diagnostic bible, formally expanded this to 106 conditions. DSM-II, published in 1968, listed 182 conditions; DSM-III (1980) listed 265 conditions; and DSM-IV (1994) listed 297. DSM-V is expected to be released later this year and will have many changes, including an unknown number of new conditions. Today, around 40 million Americans are thought to be suffering from a mental illness. In 1975, only about 25 percent of psychiatric patients received a prescription, but today almost all do, and many receive multiple drugs. The use of these drugs has spread so rapidly that levels of common antidepressants, such as Prozac, have been detected in the U.S. public water supply.

Two key studies from the 1970s illustrate some of the subjectivity underlying our psychiatric diagnoses. In 1973, David Rosenhan described an experiment in which eight healthy people who simulated fake auditory hallucinations and went for psychiatric evaluations were all hospitalized (for an average of nineteen days) and forced to agree to take antipsychotic drugs before their release. This by itself is not surprising, since doctors tend to trust patients' description of symptoms. But what was revealing was the second part of the experiment, in which a psychiatric hospital challenged Rosenhan to repeat it using its facility and Rosenhan agreed. In the subsequent weeks, the hospital's psychiatrist identified 19 (of 193) presenting patients as potential pseudopatients, when in fact Rosenhan had sent no one to the hospital at all.

In another study, in 1971, 146 American psychiatrists and 205 British psychiatrists were asked to watch videotapes of patients. In one case, involving hysterical paralysis of one arm, mood swings, and alcohol abuse, 69 percent of the Americans diagnosed schizophrenia but only 2 percent of the British did so.

Despite the DSM having been developed by many of the world's leading minds with the best of intentions, the dilemmas illustrated by these studies remain a challenge: overlapping criteria of many disorders, wide symptom fluctuations, spontaneous remission of symptoms, subjective thresholds for severity and duration, and diagnostic variations even among Anglo-Saxon cultures.

DSM-III and IV, with their translation into multiple languages, resulted in the globalization of these American diagnostic criteria, even though they were never intended as a cross-cultural export. Many foreign psychiatrists who attended the American Psychiatric Association's annual meeting began implementing these ideas in their native countries. Western pharmaceutical companies seeking new markets in emerging countries were quick to follow with large-scale campaigns marketing their new pills for newly classified mental disorders, without fully appreciating the crosscultural variations. Rates of U.S.-defined psychiatric disorders are rising in many countries, including emerging nations.

In his insightful book *Crazy Like Us: The Globalization of the American Psyche*, Ethan Watters raises the worry that by exporting an American view of mental disorders as solid scientific entities treatable by trusted pharmaceuticals, we may be inadvertently increasing the spread of such diseases. We assume that people around the world react the same way to stress as we do. We assume that mental illness around the world manifests the same way as it does in the U.S. We assume that our methods and pills are better ways to manage mental illnesses than local and traditional methods. But are these assumptions correct?

Suffering and sadness in many Asian cultures has traditionally been seen as part of a process of spiritual growth and resilience. People in other cultures react to stress differently from us. Even severe illnesses such as schizophrenia may manifest differently outside the U.S., due to cultural adaptations or degrees of social support. For example, a landmark World Health Organization study of 1,379 patients from ten countries showed that two-year outcomes of first-episode schizophrenics were much better for the patients in the poor countries than in the U.S., despite a higher proportion of American patients on medications. In my own travels to India, I have seen these trends in full bloom. As the Asian psyche becomes more Americanized, people from Bombay to Beijing are increasingly turning to pills for stress, insomnia, and depression. Is this the best direction for the entire world to follow?

SOCIAL MEDIA: THE MORE TOGETHER, THE MORE ALONE

MARCEL KINSBOURNE

Pediatric neurologist; neuroscientist; professor of psychology, The New School; coauthor (with Paula J. Kaplan), Children's Learning and Attention Problems

Mark Twain explained that good intentions lead to good consequences half the time. They lead to bad consequences half the time, also. Were he around today, he would agree that the social media are obviously well intentioned and that their good consequences need no further discussion. However, Twain would realize that he left out an even more common outcome of good intentions: good and bad consequences at the same time. I will discuss what I fear we lose when interaction between people shrinks to a disembodied voice, and worse, a throwaway electronic twittering of words and acronyms.

The sharing of information is the least of what people do when they speak with each other. Far more, speech is a rehearsal of what the listener already knows or has no interest in knowing (technically, phatic speech). Yet people all over the world seek out phatic conversation for its own sake, and research has shown that after a chat, however vacuous, the participants not only feel better but also feel better about each other. How does that come to pass?

Protracted face-to-face interaction is one of the few human behaviors not seen even in rudiment among other animal species. Its evolutionary advantage is as a mechanism for bonding—parent with child, partners with each other. The entrainment into amicable conversation implements the bonding; eye contact, attention to facial expression (smile? smile fading?), and an automatic entrainment of body rhythms, a matching of speech intonation, unconscious mimicry of each other's postures and gestures, all well documented, which is underwritten by an outpouring of oxytocin. Vigilant anticipation of the other's body language and continual adjustments of one's own demeanor in response make for an outcome of a higher order, aptly called "intersubjectivity" or "extended mind." Minds previously each preoccupied with their own concerns defer to the other's topic of interest, so as to arrive at a more shared and unified perspective on the object of attention or the topic of debate. Indeed, the harmony goes beyond the concrete and the conceptual. It ranges into the emotions; insistent bleak ruminations diffuse and scatter as the mind mingles with the mind of an intimate or congenial companion.

How to transcend the two-dimensional talking heads staring hopefully from their television perches into the void? Key to the enriched face-to-face communication is that it happens up close and personal, within arm's length, so that mutual touch is possible. The impoverished response elicited by a posting of good or bad news on Facebook or Twitter—OMG, CUTE!, how cool!, etc.—squanders the opportunity for interpersonal warmth, an embrace, admiring a ring, a diploma, a baby, and the attending reminiscences, the anticipations and sharing of feelings in depth, that happen when two people face each other, lock eye contact, and experience the reality of each other. Evolution notoriously has no foresight, nor is it embodied, but if it were, it would be spinning like a top in its grave as LOL supplants the joy of present laughter. What a waste!

Staples of contemporary interaction have the effect, perhaps intended, of distancing and alienating: the elevated podium, the massive executive desk that separates the speaker and the spoken-to, the dimmed lights as PowerPoint takes over from the human speaker and no one looks at anyone else. As for the written word, text and prose become inert "content," and the healer is relegated to insurance "provider" status in the mercantile jargon of the day. Then come the online courses, capped by online diplomas, absent any person-to-person teaching. There is no opportunity for the student to participate in an instructor's inspiration with an excited interpersonal to-and-fro that sharpens the understanding of both parties of the idea at issue. The speaker (writer) and the listener (reader) both face a void. No one is there to talk with.

Humans can do better than that. The entrainment of body rhythms has consolidated the loyalty of, and the shared mindset of, groups perhaps since times prehistoric: the rhythm of synchronous marching feet, the ritual chant, the dance around the cauldron, complete with boiled missionary. Between individuals, the salute—and, adding touch to the multimodal interpersonal mix, the handshake, the high five, couples dancing. Whether the objective is to promote military discipline, adherence to a sect or religious persuasion, or faith-healing, rhythmically entrained behavior is the catalyst and the glue. When the goal is to inculcate irrational and counterintuitive beliefs, this is more successfully done by rhythm and group ritual than by strenuous verbal persuasion. Of such is the collective, otherwise known as the in-group, for better or for worse.

I fear that weakening, if not completely relinquishing, the compelling attraction inherent in entraining both physically and mentally with others will leave people alone even as their tally of “friends” increases. I also fear that the flight from closeness to “individualism” will foster ever more vehement and destructive untrammeled self-expression, as the hobgoblins emerge from the shadows into the spotlight of the Internet and into explosive mayhem in the real world.

INTERNET DRIVEL

DAVID GELERNTER

Computer scientist, Yale University; chief scientist, Mirror Worlds Technologies; author, America-Lite: How Imperial Academia Dismantled our Culture (and Ushered in the Obamacrats)

If we have a million photos, we tend to value each one less than if we had only ten. The Internet forces a general devaluation of the written word—a global deflation in the average word's value on many axes. As each word tends to get less reading time and attention and be worth less money at the consumer end, it naturally tends to absorb less writing time and editorial attention on the production side. Gradually, as the time invested by the average writer *and* the average reader in the average sentence falls, society's ability to communicate in writing decays. And this threat to our ability to read and write is a slow-motion body blow to science, scholarship, the arts—to nearly everything, in fact, that is distinctively human, that muskrats and dolphins can't do just as well or better.

The Internet's insatiable demand for words creates global deflation in the value of words. The Internet's capacity for distributing words near instantly means that—with no lag time between writing and publication, publication and worldwide availability—pressure builds on the writer to produce more. Global deflation in the value of words creates pressure, in turn, to downplay or eliminate editing and self-editing. When I tell my students not to turn in first drafts, I sometimes have to explain, nowadays, what a first draft is.

Personal letters have traditionally been an important literary medium. The collected letters of a Madame de Sévigné, Vincent van Gogh, Jane Austen, E. B. White, and a thousand others are classics of Western literature. Why have no (or not many) "collected e-mails" been published, on paper or online? It's not only that e-mail writing is quick and casual; even more, it's the fact that we pay so little attention to the e-mails we get. Probably there are many writers out there whose e-mails are worth collecting. But it's unlikely that anyone will ever notice. And since e-mail has, of course, demolished the traditional personal letter, a major literary genre is on its last legs.

Writing ability is hard to measure, but we can try, and the news is not good. Recently the London *Daily Mail* reported on yet another depressing evaluation of American students: "While students are much more likely to call themselves gifted in writing abilities [the study concluded], objective test scores actually show that their writing abilities are far less than those of their 1960s counterparts."

It's hard to know how to isolate the effects of net-driven word devaluation in the toxic mix that our schools force-feed our children every day. But at any rate, the Internet drivel factor can't be good—and is almost certain to grow in importance as the world fills gradually with people who have spent their whole lives glued to their iToys.

At the *Huffington Post*, the future is now; the *Weekly Standard* has republished parts of a Huff-and-Puffington piece by the actor Sean Penn. Even assuming that Sean Penn is a lot more illiterate than most people, the *Post* is a respectable site, and the Penn piece is eye-opening:

> The conflicted principle here, is that which all too
> often defines and limits our pride as Americans who, in

> deference to an omnipresent filter of mono-culturalism, isolationism and division, are consistently prone toward behaviors and words, as insensitive and disrespectful, while at foremost counterproductive for the generations of young Americans who will follow us.

The only problem with this passage is that it is gibberish. The average ten-year-old hasn't fallen this far yet. But the threat is real, is way under the radar, and is likely to stay there. Prognosis: grim.

OBJECTS OF DESIRE

SHERRY TURKLE

Abby Rockefeller Mauzé Professor of the Social Studies of Science & Technology, MIT; director, MIT Initiative on Technology & Self; author, Alone Together: Why We Expect More from Technology and Less from Each Other

Children watch their parents play with shiny technical objects all day. Parents cradle them, caress them, never let them out of their hands. When mothers breastfeed their infants, the shiny objects are in their hands, at their ears. When parents bring their toddlers to the park, they share their attention with the shiny objects to the point that children are jealous—and, indeed, often go unattended. Playground accidents are up.

As soon as children are old enough to express their desires, they want the objects as well, and few parents say no. In parental slang, it has become known as the "passback"—passing back the iPhone to quiet your toddler in the car's backseat.

It has always been thus: In every culture, children want the objects of grown-up desire. And so the little shiny screens pass into playpens and cribs and then to the playground. Phones, pads, tablets, computers take the place of building blocks and modeling clay and books and dolls. The screens are interactive, scintillating, quite beautiful. They support an infinite array of simulations and worlds. Beyond interactivity, they offer connection with others. Of course, they're marketed not just as fun but as objects of artistic creation and educational worth. They may be all of these. What we know for sure is that they are deeply compelling.

The screens make children three magical promises that seem like gifts from the fairies. You will always be heard. You can put your attention wherever you want it to be. And you will never have to be alone. From the youngest age, there is a social media account that will welcome you. From the youngest age, there is a place where you can be an authority, even an authority who can berate and bully. And there is never, ever a moment when you have to quiet yourself and listen only to your inner voice. You can always find other voices.

We are embarking on a giant experiment in which our children are the human subjects. There is much that is exciting, thrilling, here. But I have some misgivings. These objects take children away from many things that we know from generations of experience are most nurturant for them. In the first instance, children are taken away from the human face and voice, because people are tempted to let the shiny screens read to children, amuse children, play games with children. And they take children away from one another. They allow children to have experiences (texting, i-chatting—indeed, talking to online characters) that offer the illusion of companionship without the demands of friendship, including the responsibilities of friendship. So there is bullying and harassment when you thought you had a friend. And there is quick, false intimacy that seems like relationship without risk, because you can always disconnect or leave the "chat."

Children become drawn in by the three promises, but they may lose out in the end. Why? Because talking to technology, or talking to others through technology, leads children to substitute mere connection for the complexities and nuance of developing conversation. Indeed, many children end up afraid of conversation. In my studies of children and technology, when I ask children "What's wrong with conversation?" they are able to answer by about age ten. To paraphrase their bottom line, "It takes place in real time and you can't control what you're going to say." They're right. That's what is wrong with conversation. And of course, particularly for a child growing up, that's what is so profoundly *right* with conversation. Children need practice dealing with other people. With people, practice never leads to perfect. But perfect isn't the goal. Perfect is the goal only in a simulation. Children become fearful of not being in control in a domain where control is not the point.

Beyond this, children use conversations with one another to learn how to have conversations with themselves. For children growing up, the capacity for self-reflection is the bedrock of development. I worry that the holding power of the screen does not encourage this. It jams that inner voice by offering continual interactivity or continual connection. Unlike time with a book, where one's mind can wander and there is no constraint on time out for self-reflection, "apps" bring children back to the task at hand just when a child's mind should be allowed to wander. So in addition to taking children away from conversation with other children, too much time with screens can take children away from themselves. It is one thing for adults to choose distraction over self-reflection. But children need to learn to hear their own voices.

One of the things modeling clay and paints and blocks did for children was slow them down. When you watch children play with them, you see how the physicality of the materials offers a resistance that gives children time to think, to use their imaginations, to make up their own worlds. Children learn to do this alone, learning to experience this time alone as pleasurable solitude for getting to know themselves. This capacity for solitude will stand them in good stead for the rest of their lives. It is in this area that I have my greatest misgiving: The screens promise that you will never have to be alone. We can already see that so many adults are terrified to be alone. At a red light or a supermarket checkout, they panic and reach for a device. Our lives with screens seem to have left us with the need to constantly connect. Instead of being able to use time alone to think, we think only of filling the time with connection.

Why is solitude so important, and why do we want to cultivate it in the young? Solitude is a precondition for creativity, but it is also where we find ourselves so that we can reach out and have relationships with other people, in which we really appreciate them as other people. So solitude is a precondition for conversation. If we aren't able to be alone with ourselves, we are at risk of using other people as "spare parts" to support our fragile selves. One of the great tasks of childhood is to develop the capacity for this kind of healthy solitude. It is what will enable children to develop friendships of mutuality and respect.

Thus my worry for kindergarten-tech: The shiny objects of the digital world encourage a sensibility of constant connection, constant distraction, and never-aloneness. And if you give them to the youngest children, they will encourage that sensibility from the earliest days. This is a way of thinking that goes counter to what we currently believe is good for children: a capacity for independent play, the importance of cultivating the imagination, essentially developing a love of solitude because it will nurture creativity and relationship.

In our still recent infatuation with our mobile devices, we seem to think that if we are connected we'll never be lonely. But in fact the truth is quite the opposite. If all we do is compulsively connect, we will be more lonely. And if we don't teach our children to be alone, they will know only how to be lonely.

I worry that we have yet to have a conversation about what seems to be a developing "new normal": the presence of screens in the playroom and kindergarten. When something becomes the new normal, it becomes hard to talk about, because it seems like second nature. But it's time to talk about what we want childhood to accomplish.

INCOMPETENT SYSTEMS

JOHN NAUGHTON
Academic; journalist; vice-president, Wolfson College, Cambridge, U.K.; author, From Gutenberg to Zuckerberg: What You Really Need to Know About the Internet

What worries me is that we are increasingly enmeshed in incompetent systems—that is, systems that exhibit pathological behavior but can't fix themselves. This is because solving the problems of such a system would require coordinated action by significant components of the system, but engaging in such action(s) is not in the short-term interest of any individual component (and may indeed be counter to its interests). So, in the end, pathological system behavior continues until catastrophe ensues.

A case study of an incompetent system is our intellectual property regime, a large part of which is concerned with copying and the regulation thereof. This regime was shaped in an analog world—in other words, an era in which copying was difficult, degenerative, and costly and in which dissemination of copies was difficult and expensive.

We now live in a digital world, in which copying is not only effortless, nondegenerative, and effectively free but is actually intrinsic to digital technology. What is a computer, after all, but a copying machine? Copying is to digital technology as breathing is to animal life; you can't have one without the other. So trying to apply an IP regime designed for analog circumstances to a world in which all media and cultural artifacts are digital offends against common sense.

Everybody knows this, but the prospects of getting a solution to the problem are poor. Why? Because moving to a more rational IP regime would require concerted action by powerful vested interests, each of which has a stake in the status quo. They're not going to move—which is why our IP regime is an incompetent system.

Even more worrying is the suspicion that liberal democracy, as currently practiced around the world, itself has become an incompetent system. The dysfunctional nature of legislative bodies, the banking and subsequent sovereign debt crisis, has revealed that the incompetence of democracies is a widespread problem. The inability—and unwillingness—of many Western governments to regulate their banks, coupled with the huge costs then unilaterally imposed on citizens to rescue commercial enterprises judged "too big to fail" has led to a widespread loss of trust in governments and a perception that even nominally "representative" democracies no longer produce administrations that serve the interests of their citizens.

DEMOCRACY IS LIKE THE APPENDIX

DYLAN EVANS

Founder & chief visionary officer, Projection Point; author, Risk Intelligence: How to Live with Uncertainty

Many people worry that there is not enough democracy in the world; I worry that we might never go beyond democracy.

In an influential essay published in 1989 and in a subsequent book, Francis Fukuyama claimed that liberal democracy was the final form of human government, the "end point of mankind's ideological evolution." Every country would eventually become democratic, and there would be no fundamental change in political organization from then on. This would be a shame, because there may be better forms of political organization that we can aspire to. But the spread of democracy may actually make it harder to discover these alternatives. To see why, we need to understand something that may at first appear counterintuitive. Democracy doesn't give most people what they want; in fact, it leads to majority dissatisfaction. When Barack Obama won the 2012 election with 51 percent of the vote, for example, it wasn't just the 49 percent who voted against him who were unhappy with the result. Most of those who voted for Obama were pretty disappointed too—because Obama ran on a platform that did not represent the ideal policy bundle of more than just a few voters. In every election, voters are forced to choose between a tiny selection of candidates, none of whom they particularly like. Everyone will be disappointed no matter which candidate wins, because nobody had a chance to vote for their ideal manifesto to begin with.

The reason for this lack of choice lies in the tendency of political parties to converge toward almost identical positions. This is a pervasive feature of modern democracies and tends to anchor society in the political middle ground. The resulting social stability has obvious advantages, in that it helps guard against political extremism. But it has less understood disadvantages, too. In particular, it hinders the development of better political systems.

Societies are complex systems, and like all such systems they can sometimes get stuck in suboptimal states. In biological systems, too, bad designs can persist despite their obvious disadvantages. A good example is the appendix. This organ used to play a part in our ancestors' digestive process, but now it is useless and we'd be better off without it. It not only does us no good but also occasionally does harm. Hundreds of thousands of people are hospitalized each year for appendicitis in the United States alone, and several hundred die from it. So why hasn't natural selection eliminated the appendix? Why does it still exist?

One intriguing suggestion, put forward by the evolutionary biologists Randolph Nesse and George Williams, is that the appendix persists because individuals with a smaller and thinner appendix are more vulnerable to appendicitis. So the normal tendency for useless organs to atrophy away to nothing is blocked, in the case of the appendix, by natural selection itself. Perhaps this idea will turn out not to be correct, but it does illustrate how the persistence of something can conceivably be explained by the very factors that make it disadvantageous.

Democracy is like the appendix. The very thing that makes majority dissatisfaction inevitable in a democracy—the voting mechanism—also makes it hard for a better political system to develop. The reforms that would be necessary to pave the way for alternative systems of governance lie well outside the safe middle ground of the median voter. Politicians advocating such reforms are unlikely, therefore, to be voted into office.

For example, one route to discovering alternative forms of governance may begin with the secession of a few cities from their parent nations, or in the creation of new cities from scratch, operating under rules different from those in the rest of the country. It's hard to imagine elected politicians getting away with such things, however, even if they wanted to. The only historical precedents so far have occurred in autocratic regimes, where leaders do not have to worry about reelection. The wave of special economic zones in China in the 1980s, beginning with Shenzhen, was driven by a small cadre of unelected officials headed by Deng Xiaoping.

I think we should worry that democracy may turn out to be a historical cul-de-sac, a place that looks pleasant enough from far away but doesn't lead anywhere better.

THE IS-OUGHT FALLACY OF SCIENCE AND MORALITY

MICHAEL SHERMER

Publisher, Skeptic *magazine; monthly columnist,* Scientific American*; author,* The Believing Brain

Ever since the philosophers David Hume and G. E. Moore identified the *is-ought* problem—the discrepancy between descriptive statements (the way something is) and prescriptive statements (the way something ought to be), most scientists have conceded the high ground of determining human values, morals, and ethics to philosophers, agreeing that science can only describe the way things are but never tell us how they ought to be. This is a mistake.

We should be worried that scientists have given up the search for determining right and wrong and which values lead to human flourishing, just as the research tools for doing so are emerging from such fields as evolutionary ethics, experimental ethics, neuroethics, and related fields. The is-ought problem (sometimes rendered as the naturalistic fallacy) is itself a fallacy. Morals and values *must* be based on the way things are, in order to establish the best conditions for human flourishing. Before we abandon the ship just as it leaves port, let's give science a chance to steer a course toward a destination where scientists at least have a voice in the conversation on how best we should live.

We begin with the individual organism as the primary unit of biology and society, because the organism is the principal target of natural selection and social evolution. Thus the survival and flourishing of the individual organism—people, in this context—*is* the basis of establishing values and morals, so determining the conditions by which humans best flourish *ought to be* the goal of a science of morality. The constitutions of human societies ought to be built on the constitution of human nature, and science is the best tool we have for understanding our nature. For example:

- We know from behavioral genetics that 40 to 50 percent of the variance among people in temperament, personality, and many political, economic, and social preferences is inherited.
- We know from evolutionary theory that the principle of reciprocal altruism—I'll scratch your back if you'll scratch mine—is universal; people do not by nature give generously unless they receive something in return.
- We know from evolutionary psychology that the principle of moralistic punishment—I'll punish you if you do not scratch my back after I have scratched yours—is universal; people do not long tolerate free riders who continually take but never give.
- We know from behavioral game theory about within-group amity and between-group enmity, wherein the heuristic is to trust in-group members until they prove otherwise to be distrustful, and to distrust out-group members until they prove otherwise to be trustful.
- We know from behavioral economics about the almost universal desire of people to trade with one another, and that trade establishes trust between strangers and lowers between-group enmity, and also produces greater prosperity for both trading partners.

These are just a few lines of evidence from many different fields of science that help us establish the best way for humans to flourish. We can ground human values and morals not just in philosophical principles—such as Aristotle's virtue ethics, Kant's categorical imperative, Mill's utilitarianism, or Rawls' fairness ethics—but in science as well. Consider the following example of how science can determine human values.

Question: What is the best form of governance for large modern human societies?

Answer: a liberal democracy with a market economy.

Evidence: liberal democracies with market economies are more prosperous, more peaceful, and fairer than any other form of governance tried.

Data: In their book *Triangulating Peace*, the political scientists Bruce Russett and John Oneal employed a multiple logistic regression model on data from the Correlates of War Project that recorded 2,300 militarized interstate disputes between 1816 and 2001. Assigning each country a democracy score between 1 and 10 (based on the Polity Project that measures how competitive its political process is, how openly leaders are chosen, how many constraints on a leader's power are in place, etc.), Russett and Oneal found that when two countries are fully democratic, disputes between them decrease by 50 percent, but when the less democratic member of a country pair is a full autocracy, it doubles the chance of a quarrel between them.

When you add a market economy into the equation, it decreases violence and increases peace significantly. For every pair of at-risk nations for which Russett and Oneal entered the amount of trade (as a proportion of GDP), they found that countries depending more on trade in a given year were less likely to have a militarized dispute in the subsequent year—controlling for democracy, power ratio, great-power status, and economic growth. So they found that democratic peace happens only when both members of a pair are democratic, but that trade works when either member of the pair has a market economy.

Finally, the third vertex of Russett and Oneal's triangle of peace is membership in the international community, a proxy for transparency. The social scientists counted the number of IGOs (Intergovernmental Organizations) that every pair of nations jointly belonged to and ran a regression analysis with democracy and trade scores, discovering that democracy favors peace, trade favors peace, and membership in IGOs favors peace, and that a pair of countries in the top tenth of the scale on all three variables are 83 percent less likely than an average pair of countries to have a militarized dispute in a given year.

The point of this exercise, as further developed by Harvard psychologist Steven Pinker, is that in addition to philosophical arguments we can make a *scientific case* for liberal democracy and market economies as a means of increasing human survival and flourishing. We can measure the effects quantitatively and from that derive science-based values that demonstrate conclusively that this form of governance is *really* better than, say, autocracies or theocracies. Scholars may dispute the data or debate the evidence, but my point is that in addition to philosophers, scientists should have a voice in determining human values and morals.

WHAT IS A GOOD LIFE?

DAVID CHRISTIAN
Professor of history, Macquarie University, Sydney; author, Maps of Time: An Introduction to Big History

The 2013 *Edge* Question invites us to identify issues that are not on the public radar but should be—questions that ought to be hot topics in schools, homes, parliaments, in the media, in the U.N. Here's an old, old question that has dropped off the radar: What is a "good life?" Providing everyone with the foundations for a good life is a basic goal of public policy. Yet how little public debate there is about the real meaning of a good life.

Perhaps we resist these questions because they conjure images of male Greek philosophers drinking watered-down wine at symposia. Can their answers offer us anything useful today? Or perhaps we think we already know the answer. And indeed we do have an easy, plausible, and powerful answer that was not available to the Greeks: Modern technologies let us imagine a world of ever increasing material abundance for everyone. Standard indices of material wealth such as GDP or GNP suggest that we live much better than the Greeks. Between 1500 and 2000 (according to the widely used estimates of the late British economist Angus Maddison), global GDP multiplied almost 150 times and per capita GDP about 10 times. This is what we call "growth." And most of us, most of the time, are happy to take "growth" as a surrogate for "the good life," which is why most politicians, economists, and entrepreneurs spend most of their time working to sustain growth.

The story of growth dominates thinking about the good life

partly because, as the Chinese say of Mao Zedong, it is 70 percent (well, perhaps 50 percent) true. To enjoy a good life, we need food, security, and protection from the elements, and we must use energy and resources to provide these goods. Attempts by psychologists to measure subjective well-being support the obvious conclusion that raising consumption levels above the poverty line is fundamental to our sense of well-being and contentment. A basic minimum of material consumption is the indispensable foundation for a good life.

Yet the story of "growth" is also at least 50 percent wrong. It is wrong in two important ways: It offers an impoverished understanding of the good life and it is steering us toward ecological chaos.

We all know that beyond a certain level (and that level may not be very high), well-being depends less and less on material consumption. If you've just had a great meal, you won't increase your well-being by immediately eating five more; restraint is a source of well-being as well as consumption. Indeed, many components of the good life do not require more consumption, because they are renewable resources. They include friendship, empathy, kindness and generosity, good conversation, a sense of beauty, a sense of physical well-being and security, a sense of contentment, a sense of intimacy, a sense of humor, and (*Edge*'s forte) a delight in good ideas. Measures of increasing consumption cannot capture these psychic goods. In March 1968, not long before he was assassinated, Robert Kennedy said in a speech at the University of Kansas:

> Our gross national product . . . counts air pollution and cigarette advertising, and ambulances to clear our highways of carnage. . . . It counts the destruction of our redwoods and the loss of our natural wonder in chaotic sprawl. . . . Yet the gross national product does not allow for the health of

> our children, the quality of their education, or the joy of their play. It does not include the beauty of our poetry or . . . the intelligence of our public debate or the integrity of our public officials. . . . [I]t measures everything, in short, except that which makes life worthwhile.

Even worse, the story of growth steers us toward ecological chaos. The biosphere is rich in resources and extraordinarily resilient, but there are limits. And what "growth" really means is ever increasing consumption of the energy and resources of the biosphere by one species, our own. Today we are learning that "growth" is pushing the biosphere to its limits. There is a real danger that biospheric systems will start breaking down, perhaps violently and fast, because we are messing with ancient, complex, unpredictable, and global metabolic pathways, such as the carbon and nitrogen cycles. If the resources of the biosphere are limited, then "growth" cannot continue indefinitely. So we have to start imagining what a good life will look like in a world of limited resources.

A basic level of material abundance is indeed the foundation. A good society will be one in which everyone enjoys the material foundations for a good life. But beyond that level, we will need to distinguish more clearly between the renewable and the nonrenewable components of a good life. Can we learn better how to appreciate and enjoy the renewable resources of a good life?

Developing a more realistic story about the good life will be an essential step toward a better life and a more sustainable society. These conversations will be complex and difficult. They will engage educators, scientists, economists, politicians, artists, entrepreneurs, and citizens as well as philosophers. But we desperately need the debate as we try to imagine a better future for our children and their children.

A WORLD WITHOUT GROWTH?

SATYAJIT DAS
Expert, financial derivatives & risk; author, Extreme Money: The Masters of the Universe and the Cult of Risk

Arthur Miller wrote that "an era can be said to end when its basic illusions are exhausted." Economic growth, the central illusion of the age of capital, may be ending.

Growth underpins every aspect of modern society. Economic growth has become the universal solution for all political, social, and economic problems, from improving living standards and reducing poverty to, now, solving the problems of overindebted individuals, businesses, and nations.

All brands of politics and economics are deeply rooted in the idea of robust economic growth, combined with the belief that governments and central bankers can exert substantial control over the economy to bring this about. In his 1925 novel *The Great Gatsby*, F. Scott Fitzgerald identified this fatal attraction: "Gatsby believed in the green light, the orgiastic future that year by year recedes before us. It eluded us then, but that's no matter—tomorrow we will run faster, stretch out our arms farther."

In reality, economic growth is a relatively recent phenomenon. It took approximately five centuries (from 1300 to 1800) for the standard of living, measured in terms of income per capita, to double. Between 1800 and 1900, it doubled again. The 20th century saw rapid improvements in living standards, which increased between five and six times. Living standards doubled between 1929 and 1957 (twenty-eight years) and again between 1957 and 1988 (thirty-one years). Between 1500 and 1820, economic production increased by less than 2 percent per century. Between 1820 and 1900, economic production roughly doubled. Between 1901 and 2000, economic production increased by a factor of something like four times.

Over the last thirty years, a significant proportion of economic growth and the wealth created relied on *financialization*. As traditional drivers of economic growth such as population increases, new markets, innovation, and increases in productivity waned, debt-driven consumption became the tool of generating economic growth. But this process requires ever increasing levels of debt. By 2008, $4 to $5 of debt was required to create $1 of growth. China now needs $6 to $8 of credit to generate $1 of growth, an increase from around $1 to $2 of credit for every $1 of growth a decade ago. Debt allows society to borrow from the future. It accelerates consumption, as debt is used to purchase something today against the promise of paying back the borrowing in the future. Growth is artificially increased by spending that would have taken place normally over a period of years being accelerated because of the availability of cheap money. With borrowing levels now unsustainable, debt-engineered growth may be at an end.

Growth was also based on policies that led to the unsustainable degradation of the environment. It was based upon the uneconomic, profligate use of mispriced nonrenewable natural resources, such as oil and water.

The problem is the economic model itself. As former Fed chairman Paul Volcker observed on December 11, 2009: "We have another economic problem which is mixed up in this of too much consumption, too much spending relative to our capacity to invest and to export. It's involved with the financial crisis, but in a way it's more difficult than the financial crisis, because it reflects the basic structure of the economy." The simultaneous end of financially engineered growth, environmental issues, and the scarcity of essential resources now threatens the end of an unprecedented period of growth and expansion.

Policy makers may not have the necessary tools to address deep-rooted problems in current models. Revitalized Keynesian economics may not be able to arrest long-term declines in growth, as governments find themselves unable to finance themselves to maintain demand. It is not clear how, if at all, printing money or financial games can create *real* ongoing growth and wealth.

Low or no growth is not necessarily a problem. It may have positive effects—for example, on the environment or conservation of scarce resources. But current economic, political, and social systems are predicated on endless economic expansion and related improvements in living standards. Growth is needed to generate higher tax revenues, helping balance increased demand for public services and the funds needed to finance these. Growth is needed to maintain social cohesion. The prospect of improvements in living standards, however remote, limits pressure for wealth redistribution. As Henry Wallich, a former governor of the U.S. Federal Reserve, accurately diagnosed: "So long as there is growth there is hope, and that makes large income differentials tolerable."

The social and political compact within democratic societies requires economic growth and improvements in living standards. Economic stagnation increases the chance of social and political conflict. Writing in *The War of the World: Twentieth-Century Conflict and the Descent of the West*, Niall Ferguson identified the risk:

> Economic volatility matters because it tends to exacerbate social conflict. . . . [P]eriods of economic crisis create incentives for politically dominant groups to pass the

> burdens of adjustment onto others. . . . Social dislocation may also follow periods of rapid growth, since the benefits of growth are very seldom evenly distributed. . . . [I]t may be precisely the minority of winners in an upswing who are targeted for retribution in a subsequent downswing.

Politicians, policy makers and ordinary people do not want to confront the possibility of significantly lower economic growth. Like Fitzgerald's tragic hero Gatsby, the incredulous battle cry is, "Can't repeat the past? Why of course you can!" But as philosopher Michel de Montaigne noted, "How many things we regarded yesterday as articles of faith that seem to us only fables today." A recent book, Alan Weisman's *The World Without Us*, was based on a thought experiment: What would a world bereft of humans revert to? We should be worried about what a world without growth—or, at best, low and uneven rates of growth—will look like.

HUMAN POPULATION, PROSPERITY GROWTH: ONE I FEAR, ONE I DON'T

LAURENCE C. SMITH

Professor and vice-chair of geography, professor of earth & space sciences, UCLA; author, The World in 2050

If population growth is a measure of a species' success, then the 20th century was astonishingly successful for *Homo sapiens.* In just one long human lifetime, we grew our living population from 1.6 to 6.1 billion, a net addition of 4.5 billion people. Now in the 21st, we exceed 7 billion and demographic computer models, fed with national birth and death statistics from countries around the globe, reflect our slowing but still climbing growth, advancing toward 9 to 10 billion by 2050 despite falling total fertility rates in much of the world.

This worries biologists and ecologists like Stanford's Paul Ehrlich (author of the 1968 bestseller *The Population Bomb*) who have often seen exponential population growth at first succeed, then fail spectacularly. In natural ecosystems, exponential growth (often called a J-curve, owing to its sharp upward curve) is the hallmark of a boom-and-crash species. Snowshoe hares (*Lepus americanus*), for example, procreate inexhaustibly until a food shortage or disease triggers a crash, decimating not just the hares but also the more abstinent lynxes (*Lynx canadensis*) that eat them. Nature is rife with grim examples of population booms and crashes, which is why Ehrlich and others grew fearful by the 1970s, as our exponential growth charged on with no signs of stabilization in sight. This inspired some extreme birth-control measures in the developing world, such as China's one-child policy and targeted sterilization programs in India. Today "population growth" is still at the top of many people's lists of the most pressing challenges facing the world. The thought of 10 billion people sharing the planet by mid-century (some 40 percent more than today) is for them truly dreadful.

Such fear is misplaced. Not because 10 billion isn't a large number; indeed, it is huge. Total populations of other large-bodied, top-level predators (bears, for example) usually number in the tens to hundreds of thousands, not in the billions. The ubiquitous mallard duck, one of the world's commonest birds, has a global population of perhaps 30 million. The quantities of water, food, fiber, arable land, metals, hydrocarbons, and other resources needed to support 10 billion people are titanic, and the prospect of looming shortages and violent competition for such resources is the straight-line link that many people make between numeric population totals and prophecies of water shortages, famines, hydrocarbon wars, and societal collapse.

Such threats are real, but unlike natural ecosystems they aren't driven by simple head count. Instead, extreme variations in consumption, both between societies of differing means and cultures (think America vs. Afghanistan) and within societies (think rural vs. urban China), dominate contemporary natural-resource needs. This is not to say that total population doesn't matter, just that lifestyle matters even more.

Consider, for example, how the material needs for electricity, plastics, rare-earth metals, and processed food must leap by many multiples to meet the requirements of a modern urban consumer living in Shanghai, as opposed to those of an agrarian peasant toiling away in the countryside. In China alone, the massive rural-to-urban migration now under way promises a billion new urban consumers by 2050, despite China's zero population growth. Africa will have 1.2 billion, nearly a quarter of the world's urban population. My UCLA colleague Jared Diamond calculates that if everyone alive today were to adopt the current lifestyles of North Americans, Western Europeans, Japanese, and Australians, global resource consumption would rise elevenfold. It would be as if the world population suddenly rose from 7 billion to 72 billion.

That, to me, is scarier than a 40-percent increase in total population, or even the economic and social "slowing pains" of falling total fertility rates. (On a graying planet, the ratio of elderly to working-age people rises, together with strains on the social safety net and health care programs.) But unfortunately, in assuaging one fear (population growth), I've raised another (prosperity growth), which, of course, is impossible to decry. The massive rural-to-urban migration—currently some 3 million people per week, equivalent to adding another Seattle to the planet every day—has lifted hundreds of millions from relentless, grueling poverty. Who among us doesn't applaud that?

Rather than worrying about world population, the smarter focus is on the real challenge—reconciling our contradictory desires to bring modernity and prosperity to all while stabilizing the innumerable natural-resource demands that they foreshadow for our planet. Ask yourself this: What do you (the modern, educated urban dweller who is most likely reading this essay) need to give up to align your resource needs closer to those of someone cultivating rice on the Irawaddy Delta? The good news is that buying a compact house, riding a bus, or eating lower down the food chain are all easier to do than tinkering with humanity's reproduction rate. All it takes is redefining our definition of success.

THE UNDERPOPULATION BOMB

KEVIN KELLY
Editor at large, Wired*; author,* What Technology Wants

For many years, overpopulation was the ur-worry. The prospect of too many people on a finite planet stood behind common environmental worries, from pollution to global warming. Significant numbers of educated couples skipped having children at all, or no more than one, so they would do their part in preventing overpopulation. In China, having a single child was a forced decision.

While the global population of humans will continue to rise for at least another forty years, demographic trends in full force today make it clear that a much bigger existential threat lies in global underpopulation.

That worry seems preposterous at first. We've all seen the official graph of expected human population growth. A steady rising curve swells past us now at 7 billion and peaks out about 2050. The tally at the expected peak continues to be downgraded by experts; currently U.N. demographers predict 9.2 billion at the top. The peak may be off by a billion or so, but in broad sweep the chart is correct.

But curiously, the charts never show what happens on the other side of the peak. The second half is so often missing that no one even asks for it any longer. It may be because it's pretty scary news. The hidden half of the chart projects a steady downward plunge toward fewer and fewer people on the planet each year—and there is no agreement on how close to zero it can go. In fact, there's much more agreement about the peak than about how few people there will be on the planet in 100 years.

A lower global population is something many folks would celebrate. The reason it's scary is that the low will keep getting lower. All around the world, the fertility rate is dropping below replacement level, country by country, so that globally there will soon be an unsustaining population. With negative population growth, each generation produces fewer offspring, who produce fewer still, until there are none. Right now, Japan's population is way below replacement level, as is that of most of Europe, Eastern Europe, Russia, the former Soviet Republics, and some Asian countries. It goes further: Japan, Germany, and Ukraine have absolute population decline; they are already experiencing the underpopulation bomb.

The shocking news is that the developing world is not far behind. While they are above replacement level, their birthrates are dropping fast. Much of Africa, South America, the Mideast, and Iran have rapidly dropping fertility rates. The drop in fertility has recently stalled in some sub-Saharan African nations, but that's because development there has stalled. When development resumes, fertility will drop again—because fertility rates are linked to urbanity. There is a deep feedback cycle: The more technologically developed a society becomes, the fewer offspring couples will have, the easier it is for them to raise their living standards, the more that progress lowers their desire for large families. The result is the spiral of modern technological population decline—a new but now universal pattern.

All it would take to break this downward spiral is that many women living in cities all around the world decide to have more than two children in order to raise the average fertility level to 2.1 children. That means substantial numbers of couples would have to have three or four children in urban areas to make up for those with none or only one. It possibly could become fashionable to have four kids in the city. The problem is that these larger families are not happening anywhere where the population has become urban, and urbanity is now the majority mode of the population and becoming more so. Every developed country on the planet is experiencing falling birthrates. The one exception has been the United States, because of its heavy immigration, primarily because of Catholic Hispanic immigrants, but even that is changing. The most recent report shows that the birthrates of Hispanic immigrants in the U.S. is dropping faster than ever. Soon the U.S. will be on a par with the rest of the world, with plunging birthrates.

To counter this scary population implosion, Japan, Russia, and Australia pay bonuses for newborns. Singapore (with the lowest fertility rate in the world) will pay couples $5,000 for a first child and up to $18,000 for a third child—but to no avail; Singapore's rate is less than one child per woman. In the past, drastic remedies for reducing fertility rates were difficult, but they worked. Drastic remedies for increasing fertility don't seem to work, so far.

Our global population is aging. The moment of peak youth on this planet was in 1972. Ever since, the average age on Earth has been increasing each year, and there is no end in sight for the aging of the world for the next several hundred years! The world will need the young to work and pay for medical care of the previous generation, but the young will be in short supply. Mexico is aging faster than the U.S., so all those young migrant workers who seem to be a problem now will soon be in demand back home. In fact, after the peak, individual countries will race against one another to import workers, modifying immigration policies, but these individual successes and failures cancel out and won't affect the global picture. The picture for the latter half of this century will look like this: Increasing technology, cool stuff that extends human life, more older people who live longer, millions of robots, but few young people. Another way to look at the human population of 100 years from now is that we'll have the same number of over-sixty-year-olds but several billion fewer youth.

We have no experience throughout human history with declining population and rising progress (including during the Black Plague years). Some modern countries with recent population decline have experienced an initial rise in GDP because there are fewer "capitas" in the per-capita calculation, but this masks long-term diminishment. But there can always be a first time!

Here is the challenge: This is a world where every year there is a smaller audience than the year before, a smaller market for your goods or services, fewer workers to choose from, and a ballooning elder population that must be cared for. We've never seen this in modern times; our progress has always paralleled rising populations, bigger audiences, larger markets, and bigger pools of workers. It's hard to see how a declining yet aging population functions as an engine for increasing the standard of living every year. To do so would require a completely different economic system, one we are not prepared for at all right now. The challenges of a peak human population are real, but we know what we have to do; the challenges of a dwindling human population tending toward zero in a developed world are scarier because we've never been there before. It's something to worry about.

THE LOSS OF LUST

TOR NØRRETRANDERS
Science writer, consultant, lecturer, Copenhagen; author,
The Generous Man: How Helping Others Is the Sexiest Thing You Can Do

We should worry about losing lust as the guiding principle for the reproduction of our species. Throughout history, human beings, guided by instincts and intense desire, have shown great wisdom in choosing partners for reproduction. Much of the aesthetic pleasure and joy we take in contemplating other members of our species is rooted in indicators for fertility, gene quality, and immune-system compatibility. Thus our lust evinces considerable prudence.

When it comes to the number of offspring, we will have collectively stabilized the world population by mid-century through decentralized decision making. This demographic transition will result not from scientific planning but from the biological cleverness of individual couples. The stabilization means that it is ever more important that the biological preferences expressed in lust dominate reproduction, since fewer babies will be born and they will live longer.

Attempts to manage reproduction through biotechnologies and screening of eggs, sperm, partners, and embryos will interfere with the lust-dominated process. That this could mean the loss of an evolved expertise in survivability is worrisome.

Moreover, the desire to reproduce leads to the advertising of good genes and general fertility through a cultural and societal display of skills and sexiness. These are major, if not dominant, sources of the unconscious drive for creating great results in science, art, and social life. Attempts to shortcut mating preferences and the matching process through clinical control could lead not only to a loss of quality in the offspring but also to a loss of cultural fertility.

NOT ENOUGH ROBOTS

RODNEY A. BROOKS
Panasonic Professor of Robotics, emeritus, MIT; chairman & chief technical officer, Rethink Robotics; author, Flesh and Machines: How Robots Will Change Us

Many recent press stories have worried that smarter robots will take away too many jobs from people. What worries me most right now is that we will not find a way to make our robots smart enough quickly enough to take up the slack in all the jobs we will need them to do over the next few decades. If we fail to build better robots soon, then our standard of living and our life spans are at risk.

Population growth and technological advance have gone hand in hand for centuries, with one enabling the other. Over the next fifty years, the world's population growth is going to slow dramatically, and instead we are faced with a demographic shift in age profile of our population unlike anything we have seen since the Shakers—and we know what happened to them.

The one-child policy in China, now well into its second generation, has already shifted the demographics of their population in visible ways. Young married couples today are the only descendants of four parents and eight grandparents; they must steel themselves for crushing responsibilities as their loved ones age.

But there is worse from a broader perspective: China has just passed its "peak nineteen-year-old" year. Nineteen- to twenty-three-year-olds are the part of the population most drawn upon for both manufacturing and military service. The strains in China are already visible. With competition for labor increasing, we have seen working conditions and wages improve dramatically in China over the last three years. That's good, from a moral point of view, and it's good from a Western independence point of view, because companies are starting to insource manufacturing back to North America and Europe. But it also means that the tools we use to write and read these essays—and, indeed, the tools we use to run our society and almost all manufactured goods we buy at superstores—are going to get more expensive to build. We need productivity tools in the East and in the West, new forms of automation and robots to increase our manufacturing productivity. Over the next few years, we will become more and more desperate for smarter robots to work in our factories.

The demographic shifts visible in China are also playing out in Japan, Europe, and the United States. Our populations are aging rapidly—slightly more slowly in the U.S., as we, for now, have higher immigration rates than those other regions. While we worry about the solvency of our social security systems, there is a second-order effect that will exacerbate the problem and make life unpleasant for all of us lucky enough to grow older. The demographic shift under way will mean there are fewer younger people to provide services to more older people, and supply and demand will increase their labor costs. Good for nurses and elder-care workers, but it will further stretch the meager incomes of the elderly and ultimately lower their standard of living, perhaps below that of our current elderly and infirm.

This is the new frontier for robots. We are going to need lots of them to take up the slack doing the thankless and hard grunt work necessary for elder care—for example, lifting people into and out of bed, cleaning up the messes that occur, and so on—so that the younger humans can spend their time providing the social interaction and personal face time we old people are going to crave.

THAT WE WON'T MAKE USE OF THE ERROR CATASTROPHE THRESHOLD

WILLIAM McEWAN
Postdoctoral researcher, MRC Laboratory of Molecular Biology, Cambridge, U.K.

Viruses replicate near the boundary of fidelity required to successfully pass information to the next generation. I worry that we will not devise a way to push them over that boundary.

We are at a strategic disadvantage in the fight against viral infection. Our genomes are large. Most of our copying errors will be deleterious to survival. We must therefore replicate our information faithfully—about 1 error in 10^{10} nucleotides per replication cycle is the mammalian rate. Meanwhile fleet-footed RNA viruses sample 4 orders of magnitude more sequence diversity with each generation. And a virus generation is not a long time.

But error-prone replication is not without limits. There comes a point, the error catastrophe threshold, where the replication of genetic information cannot be sustained. Beyond the threshold, the ability to replicate is lost. Gently tipping viruses over this threshold has been suggested as a therapeutic intervention. And what a beautiful idea! The error catastrophe threshold defines the boundary of the heritablility of information—the boundary of life. On the other side lies the abyss.

Error catastrophe has been modeled theoretically. In support of theory, tissue culture and mouse studies sustain accelerated mutation as an antiviral strategy. But unfortunately the same drugs that induce accelerated mutation in viruses are toxic to the host. Error-inducing drugs that are specific to particular viruses are feasible but, when compared to outright inhibitors, seem a risky game.

For now, we can delight in the recent knowledge that nature has already got there. The APOBEC3 family of genes systematically introduces errors into virus genomes, tipping them over the precipice. Viruses take this threat seriously. HIV has a gene whose function is to counteract APOBEC3. Without this gene, HIV cannot replicate. Other viruses seem to have lost the battle: The human genome is littered with the remnants of extinct viruses that bear the scars of APOBEC3 activity. Perhaps accelerated mutation was a contributing factor to these viruses' demise. I fear that evolution's ingenuity—accumulated at that geologically slow rate of 10^{10} base pairs per generation—may surpass our own, unless we can grasp this beautiful idea and devise some gentle, nontoxic tipping of our own.

A FEARFUL ASYMMETRY: THE WORRYING WORLD OF A WOULD-BE SCIENCE

HELENA CRONIN

Codirector, London School of Economics' Centre for Philosophy of Natural & Social Science; author, The Ant and the Peacock: Altruism and Sexual Selection from Darwin to Today

As I enter the courtyard of the British Library, I pass under the long shadow cast by the bronze colossus of Newton. With measuring-compass in hand, he is fathoming the deepest laws of the universe. But the baleful inspiration for this monument dismays me. For he is the Newton of William Blake's famous print, depicting all that the artist abhorred in his construal of Newtonian science—desiccated rationality, soulless materialism. With thoughts of Blake, the tiger and its "fearful symmetry" come to mind. And the phrase suddenly brings into sharp perspective a nagging worry that I had not articulated. I finally pinpoint what it is that so dismays me about the dismissive views of science that I regularly encounter. It is their fearful *a*symmetry—the discrepancy between the objective status of the science and its denigration by a clamorous crowd of latter-day Blakes.

If you work on the science of human nature—in particular, sex differences—that asymmetry will be all too familiar. There is a vocal constituency of educated people—some of them scientists, even biologists; social scientists; public intellectuals; journalists—people who respect science, biology, even human biology, and who, at least ostensibly, take Darwinism to be true for all living things . . . as long as it doesn't venture into our evolved human nature.

To understand the resulting asymmetry and how worrying it is, consider first a crucial distinction between two worlds.

One is the world of the objective content of ideas—in particular, of science as a body of knowledge, of actual and possible scientific theories, their truth or falsity, their refutability, tests passed or failed, valid and invalid arguments, objections met or unanswered, crucial evidence deployed, progress achieved. Darwinian science has high status in this world of objective knowledge as perhaps (in Daniel Dennett's words) "the single best idea anyone has ever had." What's more, probably uniquely in the history of science, it is unlikely to be superseded; biology will be forever Darwinian—for natural selection, it seems, is the only mechanism that can achieve design without a designer. And the Darwinian understanding of human nature is a straightforward implication of that core insight. A remarkable feat: the first serious attempt at a science of ourselves . . . and it's most likely right.

The other world is the subjective one of mental states, the personal and the social: thoughts, beliefs, perceptions, feelings, emotions, hopes, ambitions.

Armed with this distinction, we can pin down exactly what the asymmetry is and why it is so worrying.

Generally, the public reception of a scientific theory concurs by and large with the judgment of the objective world of ideas. Not, however, in the case of the scientific understanding of our evolved human nature and, above all, male and female natures. If the arguments against the evolutionary science of human nature were conducted in the world of the objective content of ideas, there would be no contest; evolutionary theory would win hands down. But as a sociological fact in the public marketplace, it loses disastrously to its vociferous critics.

How? Because, in such encounters, the objective relationship between the science and these criticisms is turned on its head; all the fundamental asymmetries are systematically reversed.

First, the "burden of proof," the burden of argument, is transferred from the criticisms onto the science; it is Darwinism that's on trial. Meanwhile, anti–Darwinian attitudes don't have to defend themselves—they are accepted uncritically, with eager credulity and indulgent suspension of disbelief.

Second, adding insult to injury, a plethora of homemade alternatives is conjured up to fill the gap where the real science should be. What does this DIY science look like? Pseudo-methodological denunciations, where even name-calling passes for argument—essentialist, reductivist, teleological, Panglossian, determinist (all very bad) and politically incorrect (very bad indeed); the immutable "entanglement" of nature and nurture, which renders nature impenetrable, thereby freeing "pure nurture" to be discussed at length. A cavalier disregard for hard-won empirical evidence—apart from scans of brains lighting up. The magical potency of "stereotyping" (bad) and "role models" (good). A logic-defying power to work miracles on *tabula rasa* psychologies, as in "socialization" (bad—a noxious source of male/female differences) and "empowerment" (good); made-up mechanisms—multitasking, self-esteem. Complaints about evolutionary science being "controversial," which is false scientifically, though (sadly) true sociologically, because they are raising the dust and then complaining that they can't see. The science-free policy that this agenda generates is epitomized by the "women in science" lobby, which is posited on a "bias and barriers" assumption and an outright rejection of, yes, the science of sex differences.

Third, there is the impact that these meretricious views make in the public realm. They are not perceived as mere opinion. And, both psychologically and sociologically, they have a voice far more influential and persuasive than their objective status warrants. Meanwhile, it is the science that is seen as tendentious and is routinely misconstrued, maligned, and dismissed.

It seems, then, that the perspective of the two worlds reveals a bleak picture of the present. But it also reveals a far more promising vista. Understanding the distinction between the autonomous objectivity of the world of ideas and the contrasting status of the psychological and social helps us appreciate the true value of science, its progressive trajectory and its unique and enduring power. And with that in mind, I feel less worried about those fearful asymmetries.

MISPLACED WORRIES

DAN SPERBER

Social & cognitive scientist, Central European University, Budapest, & Institut Jean Nicod, Paris; coauthor (with Deirdre Wilson), Meaning and Relevance

Worrying is an investment of cognitive resources laced with emotions from the anxiety spectrum and aimed at solving some specific problem. It has its costs and benefits, and so does not worrying. Worrying for a few minutes about what to serve for dinner in order please one's guests may be a sound investment of resources. Worrying about what will happen to your soul after death is a total waste. Human ancestors and other animals with foresight may have worried only about genuine and pressing problems, such as not finding food, or being eaten. Ever since, they have become much more imaginative and have fed their imagination with rich cultural inputs; that is, for at least 40,000 years—possibly much longer—humans have also worried about improving their lot individually and collectively (sensible worries) and about the evil eye, the displeasure of dead ancestors, the purity of their blood (misplaced worries).

A new kind of misplaced worries is likely to become more and more common. The ever accelerating current scientific and technological revolution results in a flow of problems and opportunities presenting unprecedented cognitive and decisional challenges. Our ability to anticipate these problems and opportunities is swamped by their number, novelty, speed of arrival, and complexity.

Every day, for instance, we have reasons to rejoice in the new opportunities afforded by the Internet. The worry of fifteen years ago—that it would create yet another major social divide, between those with access to the Internet and those without—is so last century! Actually, no technology in human history has ever spread so far, so fast, so deep. But what about the worry that by making detailed information about every user available to companies, agencies, and governments the Internet destroys privacy and threatens freedom in much subtler ways than Orwell's Big Brother? Is this what we should worry about? Or should we focus on making sure that as much information as possible is freely accessible as widely as possible, forsaking old ideas of secrecy, and even privacy, and trusting that genuine information will overcome misinformation and that well-informed people will be less vulnerable to manipulation and control—in other words, that with much freer access to information a more radical kind of democracy is becoming possible?

Genetic engineering promises new crops, new cures, improvement of the human genome. How much should we be thrilled, how much frightened? How much and how should the development of genetic engineering itself be controlled, and by whom?

New arms of destruction—atomic, chemical, biological—are becoming more and more powerful and more and more accessible. Terrorist acts and local wars of new magnitude are likely to occur. When they do, the argument will be made even more forcefully than it was in the U.S. after 9/11 that powerful states should be given the means to try and prevent them, including means that curtail democratic rights. What should we worry most about—terrorism and wars or increased limitations to rights?

Looking further into the future: Humans will soon be living with and depending on intelligent robots. Will this develop into a new kind of master-servant dialectic, with the masters alienated by their servants? Will in fact the robots themselves evolve into masters or even into intelligent, purposeful beings with no use for humans? Are such worries sound or silly?

These are just some examples. Scientific and technical developments introduce, at a faster and faster pace, novel opportunities and risks we had not even imagined. Of course, in most cases you and I form opinions as to what we should worry about. But how confidently can we hold these opinions, pursue these worries?

What I am particularly worried about is that humans will be less and less able to appreciate what they should be worrying about and that their worries will do them more harm than good. Maybe, just as in rafting through rapids, one should try not to slow down but to optimize a trajectory one does not really control—not because safety is thereby guaranteed and the optimism is justified (the worst could still happen) but because there is no better option than hope.

THERE IS NOTHING TO WORRY ABOUT, AND THERE NEVER WAS

VIRGINIA HEFFERNAN
National correspondent, Yahoo! News

We have nothing to worry about but worry itself. This has always been true.

Marshall McLuhan said that "electric circuitry is Orientalizing the West." Let's hope that the Internet finishes that job any day now. Its next move is clearly to abolish the ghastly Western *idée fixe* that even a pixel of freedom or insight is gained in the miserable practice of worrying, in the chronic brain-cell-bruising overexercise of our lizard impulses to fight or flee.

Networked computing and digital experience has decentralized the self. It has found the hallucinatory splendor in the present moment. It has underscored our fundamental interdependence.

And in its gentle way it has also satirized the stubborn Western notion of separation. The digital revolution—with all its holy artifacts—has made manifest, jubilant, and even profitable the vision of the Buddha and Shakespeare and Wordsworth and all the world's mystics, two-bit and Stanford-chaired alike: We are one. Limiting beliefs about the past and the future destroy our lives. Only mindful acceptance of present reality can bring peace and inspire wise action.

So build a bomb shelter. Send money to people who lack it. Triple-encrypt and judiciously back up every J. Crew promotional e-mail you receive, lest Internet terrorism befall us. Hustle to keep your kids on or off the Internet, eating organic or local or nothing at all. Take these actions, or none. Just don't worry about them. There is nothing to worry about, and there never was.

WORRIES ON THE MYSTERY OF WORRY

DONALD D. HOFFMAN
Cognitive scientist, UC Irvine; author, Visual Intelligence

Worry itself may be well worth worrying about. Not just for the standard reason that there is an optimal range of worry (too much or too little is inimical to health). But rather because worry poses a question for science that—to use a distinction introduced by Noam Chomsky—is a mystery rather than a problem. The question is simple: What is worry?

A short answer is that worry is a state of anxiety coupled with uncertainty about present or anticipated problems.

This answer can be fleshed out with details of the biological correlates of worry. Briefly, when we worry and experience psychological stress, there is activation of the anterior cingulate cortex and orbito-frontal cortex of the brain, which interact with each other and with the amygdala. The amygdala interacts with the hippocampus and other subcortical structures, which in turn control the hypothalamus. The hypothalamus controls the autonomic nervous system, which governs the release of epinephrine and acetylcholine, and the HPA axis, which governs the release of cortisol into the bloodstream. In short, when we worry, an impressively complex array of interacting neural and endocrine activities engage the whole person, brain and body. The correlation between worry and this complex biological activity is systematic and predictable.

Most of us know worry in a different way—not as a complex biological activity but as an unpleasant conscious experience. The experience of worry can vary from mild (as when we worry whether or not to take an umbrella) through deeply troubling (as when we worry over choosing a treatment for cancer) to near insanity (as when Lady Macbeth exclaims "Out, damned spot! Out, I say!—One; two; why then 'tis time to do't.—Hell is murky!—Fie, my lord, fie, a soldier, and afeard?").

Worry, as a conscious experience, prompts us to react, sometimes with unproductive nervous activity, other times with productive problem solving, and occasionally with compensating humor: "I'm not afraid of death," said Woody Allen. "I just don't want to be there when it happens."

Worry, then, is Janus-faced, having two complex aspects: biology and conscious experience. To answer the question "What is worry?" requires that we unite these two aspects. And therein lies the mystery.

It's natural to assume that the neural and endocrine correlates of worry are the cause or basis of the conscious experience of worry. A scientific theory, under this assumption, should explain precisely how these biological correlates cause or give rise to the varied and nuanced conscious experiences we call worry. The problem is that there is no such theory. The mystery is that, to date, there are no remotely plausible ideas.

Which is not to say there are no ideas. There are many. Perhaps certain oscillatory patterns of activity in the right neuronal circuits, certain quantum states in the right neuronal microtubules, certain degrees of informational complexity in the right reentrant thalamo-cortical loops, or certain functional properties of the right cortical and subcortical brain areas are responsible for causing, or giving rise to, the conscious experience of worry.

Those are interesting ideas to explore. But taken in their present state as scientific theories, they bring to mind a well-known cartoon in which two geeks stand before a chalkboard on which are scribbled nasty equations, followed by the phrase "and then a miracle occurs," followed by more nasty equations. One geek says to the other, "I think you should be more explicit here in step two." Unfortunately, this is the state of each idea. A reentrant thalamo-cortical loop achieves the right informational complexity, then a miracle occurs, then the conscious experience of worry results. Microtubules enjoy the right quantum states, then a miracle occurs, then the experience of worry results. There is no explanation of how or why the conscious experience of worry appears, much less why this experience has precisely the qualities it does. No one can make the leap from biology to conscious worry without waving a magic wand at precisely the point where one had hoped for an explanation.

This worrisome predicament has prompted desperate responses. Perhaps conscious worries are an illusion; there are no such things, and we have been worrying over nothing. Or perhaps conscious worries are real and caused by biology, but evolution has not equipped us with the concepts we need to understand how; if so, then the biological provenance of worry will remain a mystery to us, at least until we chance upon a mutation that endows the needed concepts.

What *should* we be worried about? This *Edge* Question itself implies an assumption that worry has causal powers—that if, for instance, we worry about the right topics, then this worry might prompt productive problem solving.

But does the conscious experience of worry really have causal powers? If we answer yes, then we must explain how its causal powers are related to those of biology. Once again, we face an unsolved mystery. Thus many biologists answer no. They assume, that is, that the brain somehow causes conscious worries but that conscious worries, and conscious experiences more generally, cause nothing. This successfully exorcises the mystery of causation only to introduce another mystery: How and why did conscious experiences evolve? Natural selection can select only among traits that have causal consequences for fitness. If conscious experiences have no causal consequences, then *a fortiori* they have no consequences for fitness and thus are not subject to natural selection. So the experience of worry is not a product of natural selection: a worrisome conclusion.

Well, what *should* we be worried about? We should be worried about what we're really doing when we worry. Unless, of course, there are no worries or worries do nothing. If there are no worries, then there is nothing to worry about, and it makes no sense even to advise "Don't worry, be happy." If there are worries but they do nothing, then you can worry or not, as you please—it makes no difference. But the worry is that there really are worries, that worries really do something, and that what they are, how they arise, and what they can do is, for now, a mystery.

THE DISCONNECT

BARBARA STRAUCH
Science editor, New York Times*; author,* The Secret Life of the Grown-up Brain

Every day, as science editor of the *New York Times*, I think about how we convey increasingly complex science to the general public. How do we explain the apparent discovery of the Higgs boson and help our readers understand what an enormous, astonishing—even beautiful—discovery it is?

Luckily, here at the *Times* we have writers like Dennis Overbye, who not only deals every day with the science of the cosmos but who can write about it with the poetry it often deserves.

Luck is an important ingredient here. The *Times* remains committed to a deep coverage of science. But such a commitment feels increasingly lonely. Over the past several years, I have watched science and health coverage by general-interest newspapers shrink. As a health and science editor, I used to pick up other major newspapers with trepidation, knowing there might be a good story we missed or an important angle we overlooked. The *Times* had serious competition.

Today, sadly, that is often not the case. Coverage of health and science in general-interest newspapers has declined dramatically. Reporters whose work I have long admired have moved on to other things or retired or been fired, as science staffs have been slashed. True, there are many, many more good Web sites and some excellent blogs covering a wide range of science topics, but most are aimed at smaller segments of readers, who search out information focused in specific areas. Some general-interest papers in other countries continue to value science coverage, but unless you're a reader with access to those publications, that doesn't help much.

Something quite serious has been lost. And of course this has ramifications not only for the general level of scientific understanding but also for funding decisions in Washington—and even access to medical care. And it's not good for those of us at the *Times*, either. Competition makes us all better.

This decline in general-interest science coverage comes at a time of divergent directions in the general public. At one level, there seems to be increasing ignorance. After all, it's not just science news coverage that has suffered but also the teaching of science in schools. And last year we went through a political season that showed how all this can play out, with major political figures spouting off one silly statement after another, particularly about women's health (most memorably Missouri Republican Senate candidate Todd Akin: "If it's a legitimate rape, the female body has ways to try to shut that whole thing down"). Here at the *Times* we knew the scientific discourse on these topics had gotten so ridiculous—and dangerous—that we launched a team of reporters on a series we called "Political Science," with a string of stories that tried to set the scientific record straight.

But something else is going on as well. Even as there seems to be, in some pockets, an increasing ignorance of science, there is a growing interest in it. It's easy to see, from where I sit, how high that interest is. Articles about anything scientific, from the current findings in human evolution to the latest rover landing on Mars, not to mention new genetic approaches to cancer—and yes, even the Higgs boson—zoom to the top of our newspaper's most e-mailed list.

We know our readers love science and cannot get enough of it. And it's not just our readers. As the rover *Curiosity* approached Mars, people of all ages in all parts of the country had *Curiosity* parties to watch news of the landing. Mars parties! Social media, too, has shown us how much interest there is across the board, with YouTube videos and tweets on science often becoming instant megahits.

So what we have is a high interest and a lot of misinformation floating around. And we have fewer and fewer places that provide real information to a general audience that's understandable, at least by those of us who do not yet have our doctorates in astrophysics. The disconnect is what we should all be worried about.

Still, I should also take a moment to mention a few things I'm worrying less about. And this, too, is a bit of a contradiction. In some cases, in my dozen or so years in the Science Department at the *Times*, I have watched as our readers—all of us, actually—have become more sophisticated. Misunderstanding and hype have not gone away, but over the last decade we all have gained a more nuanced understanding of how our medical-industrial complex operates—and the money that often drives it. And we have a clearer understanding of the complexity of common diseases, from mental illness to heart disease to Alzheimer's. We have reached a common understanding that there is no magic bullet to fix such diseases or even address the problems in our health care system.

While we still have a long way to go, the conversation is beginning to change in these areas and others. The constant drumbeat about obesity is, as we reported recently, showing some signs of having an impact: Obesity rates are edging downward in children. We all understand, too, that medicine has gotten too expensive and that there is a lot of overtreatment going on in this country. Again, we have moved past the point of thinking there are quick solutions to these issues. But more people are talking about such topics—at least a little here and there—without so much shouting about death panels.

In a few of these areas, I am, when I think about it, oddly hopeful.

SCIENCE BY (SOCIAL) MEDIA

MICHAEL I. NORTON

Associate professor of business administration, Harvard Business School; coauthor (with Elizabeth Dunn), Happy Money: The Science of Smarter Spending

Check the "most e-mailed" lists on Web sites for periodicals ranging from the *New York Times* to FoxNews.com and you'll often see, sprinkled in with major world events and scandals, a story about a new scientific finding: "RED WINE LINKED TO LONGEVITY," or "CLIMATE CHANGE CALLED INTO QUESTION," or "EATING DIRT IS GOOD FOR YOU."

While the increasing attention given to science by the media is primarily a positive development—surely we want a scientifically literate population, and research appearing only in obscure journals will not help achieve this goal—we should be worried about the exploding trend in "science by media," for at least two reasons.

First, it is not clear that the best science is the science that gets known best. In one study that examined media coverage of research presented at a major scientific conference, fully 25 percent of the stories that appeared in the media *never* appeared in a scientific journal. That's right: Fully one quarter of the science that laypeople encountered was not solid enough to pass muster when reviewed by experts. This same trend was true for research that made the front pages of major newspapers, the stories most likely to be read.

The problem is likely exacerbated by the rise of social media: Even if we miss the initial coverage of some new scientific finding, we are now more likely to encounter it as a tweet or a post on Facebook. Worse still, social media often encourage quick, superficial engagement. We see the title—"Red Wine Linked to Longevity"—without reading further to find out, for example, the amount of red wine that might have health benefits (one ounce a day? one gallon?) and for whom (everyone? only people with red hair?).

To be clear, I am not blaming media (social or otherwise) for this problem. It's not the job of journalists or laypeople to discern good from bad science. Even scientists in closely adjoining fields are often unable to discern good from bad science in that seemingly similar discipline. If anything, scientists themselves contribute to the problem. One analysis of media coverage of scientific findings demonstrated that roughly 50 percent of news stories overemphasized the positive effects of some experimental intervention. The biggest predictor of media engagement in this kind of spin? The presence of spin in summaries of the research written by scientists themselves.

Second, as the science that laypeople encounter is presented to them more and more by the media, the biases—presumed and actual—of those outlets will likely call into question the objectivity of the science appearing there and of the scientists who conducted the research. Consider two hypothetical research findings: one suggesting that gun possession increases violent assaults ("guns bad"), and one suggesting that gun possession decreases violent assaults ("guns good"). Imagine each of these articles published in the *New York Times* and also appearing on Fox News.

Would your opinion of the soundness of the underlying science be influenced by the publication outlet? My guess is that more people would be more likely to believe a "guns good" piece in the *Times* and a "guns bad" piece on Fox News, precisely because these pieces are at such odds with the general tenor of those publications. And considering the reverse shows the larger problem: Readers may discount "guns good" research that appears on Fox News and "guns bad" research in the *New York Times* regardless of the actual quality of the underlying science.

In sum, the science that laypeople encounter will become increasingly unfiltered by scientific experts. And even when science has been vetted by experts, laypeople will increasingly make their own determination of its credibility based not on the quality of the research but on the media outlet in which it appears. And this perceived credibility will determine their likelihood of passing that science along to others via social media. This "science by (social) media" raises the curious possibility of a general public that reads more and more science while becoming less and less scientifically literate.

UNFRIENDLY PHYSICS, MONSTERS FROM THE ID, AND SELF-ORGANIZING COLLECTIVE DELUSIONS

JOHN TOOBY

Founder of evolutionary psychology; codirector, Center for Evolutionary Psychology and professor of anthropology, UC Santa Barbara

The universe is relentlessly, catastrophically dangerous, on scales that menace not just communities but civilizations and our species as well. A freakish chain of improbable accidents produced the bubble of conditions necessary for the rise of life, our species, and technological civilization. If we continue to drift obliviously inside this bubble, taking its continuation for granted, then inevitably—sooner or later—physical or human-triggered events will push us outside and we will be snuffed like a candle in a hurricane.

We are menaced by gamma-ray bursts (which scrub major regions of their galaxies free of life); nearby supernovae; asteroids and cometary impacts (which strike Jupiter every year or two); Yellowstone-like supereruptions (the Toba supereruption some 70,000 years ago was a near-extinction event for humans), civilization-collapsing coronal mass ejections (which would take down the electrical grids and electronics underlying technological civilization in a way they couldn't recover from, since their repair requires electricity supplied by the grid; this is just one example of the more general danger posed by the complex, fragile interdependence inherent in our current technology); and many other phenomena, including those unknown to us.

Here's one that no one talks about: The average G-type star shows a variability in energy output of around 4 percent. Our sun is a typical G-type star, yet its observed variability in our brief historical sample is only one-fortieth of this. When or if the sun returns to more typical variation in energy output, this will dwarf any other climate concerns.

The emergence of science as a not wholly superstitious and corrupt enterprise is slowly awakening our species to these external dangers. As the brilliant T-shirt says, an asteroid is nature's way of asking how your space program is doing. If we are lucky, we might have time to build a robust, hardened planetary and extraplanetary hypercivilization able to surmount these challenges. Such a hypercivilization would have to be immeasurably richer and more scientifically advanced to prevent, say, the next Yellowstone supereruption or buffer a 2-percent drop in the sun's energy output. (Indeed, ice ages are the real climate-based ecological disasters and civilization enders—think Europe and North America under a mile of ice). Whether we know it or not, we are in a race to forge such a hypercivilization before these blows fall. If such threats seem too distant, low-probability, or fantastical to belong to the "real" world, then let them serve as stand-ins for the much larger number of more immediately dire problems whose solutions also depend on rapid progress in science and technology.

This raises a second category of menaces—hidden, deadly, ever adapting, already here—that worries me even more: the evolved monsters from the id that we all harbor (e.g., group identity, the appetite for prestige and power, etc.), together with their disguised offspring, the self-organizing collective delusions we all participate in and mistake for reality. (As the cognoscenti know, the technical term *monsters from the id* originated in *Forbidden Planet*.) We need to map and master these monsters and the dynamics through which they generate collective delusions if our societies are to avoid near-term, internally generated failure.

For example, cooperative scientific problem solving is the most beautifully effective system for the production of reliable knowledge the world has ever seen. But the monsters haunting our collective intellectual enterprises typically turn us into idiots. Consider the cascade of collective cognitive pathologies produced in our intellectual coalitions by in-group tribalism, self-interest, prestige seeking, and moral one-upsmanship. It seems intuitive to expect that being smarter would lead people to have more accurate models of reality. On this view, intellectual elites therefore ought to have better beliefs and should guide their societies with superior knowledge. Indeed, the enterprise of science is—as an ideal—specifically devoted to improving the accuracy of beliefs. We can pinpoint where this analysis goes awry, however, when we consider the multiple functions of holding beliefs. We take for granted that the function of a belief is to be coordinated with reality, so that when actions are based on that belief, they succeed. The more often beliefs are tested against reality, the more often accurate beliefs displace inaccurate ones (e.g., through feedback from experiments, engineering tests, markets, natural selection). However, there is a second kind of function to holding a belief that affects whether people consciously or unconsciously come to embrace it—the social payoffs from being coordinated or discoordinated with others' beliefs (Socrates' execution for "failing to acknowledge the gods the city acknowledges"). The mind is designed to balance these two functions: coordinating with reality and coordinating with others. The larger the payoffs to social coordination and the less commonly beliefs are tested against reality, then the more often social demands will determine belief; network fixation of belief will predominate. Physics and chip design will have a high degree of coordination with reality, while the social sciences and climatology will have less.

Because intellectuals are densely networked in self-selecting groups whose members' prestige is linked (e.g., in disciplines, departments, theoretical schools, universities, foundations, media, political/moral movements, and other guilds), we incubate endless self-serving elite superstitions, with baleful effects: Biofuel initiatives starve millions of the planet's poorest. Economies around the world still apply epically costly Keynesian remedies despite decisive falsification of Keynesian theory by the postwar boom. (Government spending was cut by two-thirds and 10 million veterans were dumped into the labor force, while Paul Samuelson predicted "the greatest period of unemployment and industrial dislocation which any economy has ever faced.") I personally have been astonished over the last four decades by the fierce resistance of the social sciences to abandoning the blank-slate model in the face of overwhelming evidence that it is false. As physicist Richard Feynman pithily put it, "Science is the belief in the ignorance of experts."

Sciences can move at the speed of inference when individuals need only to consider logic and evidence. Yet sciences move glacially (Max Planck's "funeral by funeral") when the typical scientist, dependent for employment on a dense in-group network, has to get the majority of her guild to acknowledge fundamental embarrassing errors. To get science systematically moving at the speed of inference—the key precondition to solving our other problems—we need to design our next-generation scientific institutions to be more resistant to self-organizing collective delusions by basing them on a fuller understanding of our evolved psychology.

MYTHS ABOUT MEN

HELEN FISHER

Biological anthropologist, Rutgers University; author, Why Him? Why Her? How to Find and Keep Lasting Love

Scientists and laymen have spent the last fifty years dispelling myths about women. I worry that journalists, academics, and laymen will continue to perpetuate an equal number of myths about men. Annually, in 2010, 2011, and 2012, I have conducted a national survey of singles in collaboration with a U.S. dating service. Together we designed a questionnaire with some 150 queries (many with up to 10 sub-questions) and polled over 5,000 single men and women. We did not sample the members of the dating site; instead, we collected data on a national representative sample based on the U.S. census. All were divorced, separated, widowed, or had never married; none were engaged, living with a partner, or in a serious relationship. Included were the appropriate number of blacks, whites, Asians, and Latinos; gays, lesbians, bisexuals, and heterosexuals; rural, suburban, and urban folks from every region of the United States, ranging in age from twenty-one to seventy-plus. The data paint a different portrait of men than do America's chattering class.

Foremost, men are just about as eager to marry as women. In the 2011 sample of people in their twenties, 68 percent of the men wanted to wed, along with 71 percent of the women, and 43 percent of the men and 50 percent of the women hoped to have children. Journalists have suggested that men want children because they don't have to change diapers. But men spend a great deal of metabolic energy at child care. To support their young, men take the dangerous jobs—90 percent of people who die at work are men. Moreover, men universally confront an intruding thief and generally drive the family car through the raging blizzard.

Men aren't "players," either. When asked about their approach to dating, only 3 percent replied, "I would just like to date a lot of people." Men are as eager to find a partner as women are; indeed, men find loneliness as stressful as women do. And men are far less picky in their search. In the 2011 sample, only 21 percent of them reported that they "must have," or find it "very important" to have, a mate of their ethnic background, vs. 31 percent of women. Only 18 percent of men (as opposed to 28 percent of women) reported that they "must have," or find it "very important" to have, a partner of the same religion. The men were less interested than were the women in a partner of the same educational background and political affiliation. And 43 percent of the men in their thirties and forties would make a commitment to a woman who was ten or more years older. Women are the picky sex.

Men fall in love faster, too—perhaps because they are more visual. Men experience love at first sight more regularly than women, and men fall in love just as often. Indeed, men are just as physiologically passionate as women. When my colleagues and I have scanned men's brains using fMRI, we have found that they show as much activity as do women in neural regions linked with feelings of intense romantic love. Interestingly, in the 2011 sample we also found that when men fall in love they are faster to introduce their new partner to friends and parents, more eager to kiss in public, and want to "live together" sooner. Then, when they're settled in, men have more intimate conversations with their wives than women do with their husbands—because women have many of their intimate conversations with their girlfriends.

Last, men are just as likely as women to believe that you can stay married to the same person forever (76 percent of both sexes). And data from other studies show that after a break-up men are 2.5 times more likely to kill themselves.

In fact it is women who seek more independence in a committed relationship. Women want more personal space (77 percent of women vs. 56 percent of men). More women are reluctant to share their bank account (35 percent of women vs. 25 percent of men). Women are more eager to have girls' night out (66 percent) than men are to go out with the boys (47 percent); and women are more likely to want to vacation with their female buddies (12 percent) than men with their male buddies (8 percent).

Two questions in these annual surveys were particularly revealing: "Would you make a long-term commitment to someone who had everything you were looking for but with whom you were not in love?" And "Would you make a long-term commitment to someone who had everything you were looking for but to whom you did not feel sexually attracted?" Thirty-one percent of men were willing to form a partnership with a woman they were not in love with, as opposed to 23 percent of women. Men were also slightly more likely to enter a partnership with someone they were not sexually attracted to (21 percent of men vs. 18 percent of women). Men in their twenties were the most likely to forgo romantic and sexual attraction to a mate; the least likely were women over sixty!

Why would a young man forfeit romance and better sex to make a long-term partnership? I suspect it's the call of the wild. When a young man finds a good-looking, healthy, popular, energetic, intelligent, humorous, and charming mate, he might be predisposed to take this opportunity to breed despite the passion he might have for another woman, one he knows he would never want to wed. When the "almost right" woman comes along, the ancestral drive to pass on his DNA toward eternity trumps his sexual and romantic satisfaction with a less appropriate partner.

The sexes have much in common. When asked what they were looking for in a partnership, over 89 percent of men and women said they "must have" or find it "very important" to have a partner they can trust, in whom they can confide, and who treats them with respect. These three requirements top the list for both sexes, in all years. Gone is the traditional need to marry someone from the "right" ethnic and religious background who will fit into the extended family. Marriage has changed more in the last 50 years than in the last 10,000. Men, like women, are turning away from traditional family customs, instead seeking companionship and self-fulfillment.

In the Iliad, Homer called love "magic to make the sanest man go mad." This brain system lives in both sexes. And I believe we'll make better partnerships if we embrace the facts: Men love—just as powerfully as women.

THE MATING WARS

DAVID M. BUSS

Professor of psychology, University of Texas, Austin; coauthor (with Cindy M. Meston): Why Women Have Sex; *author,* The Dangerous Passion: Why Jealousy Is As Necessary As Love and Sex

Sexual deception, the difficulties of attracting viable marriage partners, intimate-partner violence, infidelity, mate poaching, divorce, and post-breakup stalking—these diverse phenomena are all connected by a common causal element: an unrelenting shortage of valuable mates. The dearth of desirable mates is something we should worry about, for it lies behind much human treachery and brutality.

Despite the fact that many equate evolution with survival selection, survival is important only inasmuch as it contributes to successful mating. Differential reproductive success, not differential survival success, is the engine of evolution by selection. You can survive to old age, but if you fail to mate, you fail to reproduce, and your genes bite the evolutionary dust. We are all descendants of a ruthless selective process of competition for the most valuable mates—those who can give our children good genes and an array of resources ranging from food and shelter to the social skills needed to scramble up the status hierarchy.

We're uncomfortable placing a value on other humans. It offends our sensibilities. But the unfortunate fact is that mate value is not distributed evenly. Contrary to yearnings for equality, all people simply are not equivalent in the currency of mate quality. Some are extremely valuable—fertile, healthy, sexually appealing, resource-rich, well-connected, personable, and willing and able to confer their bounty of benefits. At the other end of the distribution are those less fortunate, perhaps less healthy, with fewer material resources, or imbued with personality dispositions such as aggressiveness or emotional instability that inflict heavy relationship costs.

The competition to attract the most desirable mates is ferocious. Consequently, those most valuable are perpetually in short supply vis-à-vis the many who desire them. People who are themselves high in mate value succeed in attracting the most desirable partners. In the crude informal American metric, the 9s and 10s pair off with other 9s and 10s. And with decreasing value from the 8s to the 1s, people must lower their mating sights commensurately. Failure to do so produces a higher probability of rejection and psychological anguish. As one woman advised her male friend who bemoaned his frustration about his lack of interest in the women attracted to him and the unreciprocated interest by women to whom he was attracted, "you're an 8 looking for 9s and being sought after by 7s."

Another source of problems on the mating market comes from deception. Scientific studies of online dating profiles reveal that men and women both try to appear higher in mate value than they truly are, on precisely the dimensions valued by the opposite sex. Men exaggerate their income and status and tack on a couple of inches to their real height. Women present as ten to fifteen pounds lighter than their real weight and some shave years off of their actual age. Both show unrepresentative photos, sometimes taken many years earlier. Men and women deceive in order to attract mates at the outer limit of their value range. Sometimes they deceive themselves. Just as 94 percent of professors believe they are "above average" for their department, on the mating market many think they're hot when they're not.

Despite valiant efforts, men's attempts to increase their market value in women's eyes do not always work. Many fail. Dating anxiety can paralyze men brave in other contexts. Some spurned men become bitter and hostile toward women after repeated rejections. As Jim Morrison of The Doors once noted, "Women seem wicked when you're unwanted."

Mating difficulties do not end among those successful enough to attract a partner. Mate-value discrepancies open a Pandora's box of problems. An omnipresent challenge within romantic relationships derives from mate-value discrepancies: when an 8 mistakenly pairs up with a 6, when one member of an initially matched couple plummets in mate value, or even when one ascends more rapidly professionally than the other. Jennifer Aniston's hold on Brad Pitt proved tenuous. Mate poachers lure the higher-value partner, driving wide, initially small wedges: "He's not good enough for you;" "She doesn't treat you well;" "You deserve someone better . . . like me." Empirically, the higher-mate-value partner is more susceptible to sexual infidelity, emotional infidelity, and outright defection.

The lower-mate-value partners typically struggle mightily to prevent infidelity and breakup, using tactics ranging from vigilance to violence. Intimate-partner battering, abhorrent as it is, has a disturbing functional logic. Since self-esteem is in part a psychological adaptation designed to track one's own mate value, blows to self-esteem cause reductions in self-perceived mate value. Physical and psychological abuse predictably harm the victim's self-esteem, narrowing the perceived discrepancy between a woman's and her partner's mate value and sometimes causing her to stay with her abuser.

Those who succeed in breaking up and leaving are sometimes stalked by former partners—typically men who know or sense that they will never again be able to attract a woman as valuable as the one they have lost. Studies I've conducted in collaboration with Dr. Joshua Duntley reveal that as many as 60 percent of women and 40 percent of men have been victims of stalking. Many stalkers are sustained by the false belief that their victims truly love them but just don't realize it yet. Stalking, like intimate-partner violence, too has a disconcerting functional logic. It sometimes works in luring the woman back.

There is no easy fix for the great shortage of desirable mates. In the undemocratic world of mating, every success inevitably comes as a loss to vying rivals. Every human who conceives can be deceived. Mate poachers will always be ready to pounce. The pleasures of sexual temptation come in the here-and-now; the costs of infidelity lie in the distant and uncertain future. But perhaps a keener awareness of mate-value logic will give us the tools to curtail the more sinister products of the mating wars.

WE DON'T DO POLITICS

BRIAN ENO
Artist; composer; recording producer: U2, Coldplay, Talking Heads, Paul Simon; recording artist

Most of the smart people I know want nothing to do with politics. We avoid it like the plague—like *Edge* avoids it, in fact. Is this because we feel that politics isn't where anything significant happens? Or because we're too taken up with what we're doing, be it quantum physics or statistical genomics or generative music? Or because we're too polite to get into arguments with people? Or because we just think that things will work out fine if we let them be—that the Invisible Hand or the technosphere will mysteriously sort them out?

Whatever the reasons for our quiescence, politics is still being done—just not by us. It's politics that gave us Iraq and Afghanistan and a few hundred thousand casualties. It's politics that's bleeding the poorer nations for the debts of their former dictators. It's politics that allows special interests to run the country. It's politics that helped the banks wreck the economy. It's politics that prohibits gay marriage and stem cell research but nurtures Gaza and Guantanamo.

But *we* don't do politics. We expect other people to do it for us, and we grumble when they get it wrong. We feel that our responsibility stops at the ballot box, if we even get that far. After that, we're as laissez-faire as we can get away with.

• What worries me is that while we're laissez-ing, someone else is faire-ing.

THE BLACK HOLE OF FINANCE

SETH LLOYD
Professor of quantum mechanical engineering, MIT; author, Programming the Universe

In the fall of 2007, investment banks tottered around like blind lame giants, bleeding cash and exotic financial instruments. There followed a huge sucking sound and a queazy combination of explosion and implosion. Some banks survived, some didn't, and my retirement savings halved in value. There wasn't much to do about the money, but there was something I could do to comfort myself, and that was to construct a scientific theory of what was happening. When in worry and doubt, work it out, preferably with some fancy equations.

At the level of metaphor, the financial implosion of an investment bank resembles the formation of a supermassive black hole in the early universe. A giant star, a million times the mass of our sun, burns through its nuclear fuel in a few tens of thousands of years. After it has consumed its nuclear fuel, it can no longer generate the heat and pressure required to fend off the force of gravity. Unable to support its own weight, the star collapses. As it implodes, it blows off its outer layer in an explosion moving at the speed of light.

In contemplating the financial wreckage, I realized that the similarity between financial collapse and gravitational collapse is not merely a metaphor. In fact, it is possible to construct a mathematical theory that applies equally to gravitational collapse and financial collapse. The key ingredient is the existence of negative energy: In both Newton's and Einstein's theory of gravity, the energy in matter is positive but the energy in gravity is negative. In the universe as a whole, the positive energy of mass and kinetic energy is exactly counterbalanced by the negative energy of the gravitational field, so that the net energy of the universe is effectively zero.

The analog of energy in financial systems is money. Households, companies, governments, and of course investment banks have assets (positive money) and debts (negative money). When the companies in which I own stock send me their annual reports, I note with fascination that their assets and net obligations sum—magically—to zero. (This magical accounting might have something to do with the collapse of my retirement fund.) So let's look at theories of systems that have positive stuff and negative stuff where the total amount of stuff sums to zero.

Start with gravity. Stars, galaxies, or the universe itself, which possess both positive and negative energy, behave differently from things like cups of coffee, whose energy is entirely positive (ignoring the gravitational and psychological effects of caffeine). In particular, gravitational systems exhibit a weird effect that goes under the name of "negative specific heat." The specific heat of a cup of coffee measures how the temperature of the coffee goes down as it loses energy by radiating heat to its surroundings. As the cup radiates energy, it cools down. Bizarrely, as a star, galaxy, or cloud of interstellar dust radiates energy, it heats up: The more energy a star loses, the hotter it gets. The star has negative specific heat.

If a cup of coffee had negative specific heat, when you put it down on the counter and forgot about it for a few minutes it wouldn't cool down, it would get hotter and hotter. The longer you forgot about it, the hotter and hotter it would get, eventually exploding in a fountain of superheated coffee. Conversely, if an ice cube had negative specific heat, the more heat it absorbed, the colder it would get. If you left such an ice cube on the counter, it would absorb heat from its hotter surroundings. As it absorbed heat, it would get colder and colder, sucking more and more heat from its surroundings until it and everything around it was drawn inexorably toward absolute zero.

Negative specific heat doesn't immediately lead to a catastrophe. In a star, the hydrogen fusing into helium at temperatures of millions of degrees has positive specific heat that counterbalances the negative specific heat of gravitation, leading to a harmonious generation of lots of free energy over billions of years. Life on Earth is a product of this harmony. Once the sun burns through its supply of nuclear fuel, however, gravity will dominate and our star will collapse.

Now turn to financial systems. As in gravitational systems, the mere existence of "negative money," or debt, need not lead to collapse: Just as in a star, the interplay between generation of assets/positive energy and debt/negative energy can proceed harmoniously and produce lots of goods. But the possibility of implosion always exists. What drives a star or an investment bank over the threshold from stable generation of energy/wealth to runaway explosion and collapse? Here's where the detailed mathematical model can help.

Around 1900, the physicist Paul Ehrenfest was trying to understand how molecules bounce around in a gas. He constructed a simple model, now called the Ehrenfest urn model. Take two urns and a bunch of balls. Initially all the balls are in one of the urns. Number each of the balls, and then pick a number at random and move the ball labeled with that number to the other urn. What happens? Initially, balls tend to move from the full urn to the empty urn. As the initially empty urn fills up, balls start to move back to the other urn as well. Eventually, each urn has approximately the same number of balls. This final state is called equilibrium. In the Ehrenfest urn model, equilibrium is stable: Once the number of balls in the two urns becomes approximately equal, it stays that way, with small fluctuations due to the random nature of the process.

The mathematical model I constructed is a simple generalization of Ehrenfest's model. In my model, there are white balls (positive energy/assets) and black balls (negative energy/debt), which can be created or destroyed in pairs. As a result, the number of white balls is always equal to the number of black balls, but the total number of balls is not conserved. In this model, there are two types of processes: Balls are moved at random from urn to urn, as before, and balls can be created or destroyed in pairs within an urn.

The urn model with creation and destruction has two distinct forms of behavior. If pairs of balls are created and destroyed at the same rate in both urns, then the behavior of the system is similar to the Ehrenfest urn model: Both urns end up with roughly the same number of balls, which fluctuates up and down over time in stable equilibrium. By contrast, if the urn with more balls is allowed to create pairs at a higher rate than the urn with fewer balls, then the behavior is unstable: The urn with more balls will acquire more and more balls, both black and white. If destruction also occurs at a higher rate in the urn with more balls, then the number of balls in that urn will explode and then collapse. In physical terms, this unstable behavior comes about because allowing the "wealthier" urn to create balls at a higher rate gives the overall system negative specific heat, so that stable equilibrium is impossible.

In financial terms, creating a pair of balls is analogous to creating debt, and destroying a pair of balls is analogous to retiring debt. The urn model implies that if wealthier institutions, with more balls, can create debt at a higher rate than less wealthy institutions, then the flows of assets and debt become unstable. That is, if wealthier institutions have higher leverage, then economic equilibrium goes unstable. Lehman Brothers was leveraged at around 30 to 1 at the time of its collapse: It had been able to borrow $30 for each dollar it actually possessed. The urn model's criteria for instability were met. The signature of unstable equilibrium is that ordinary transactions, such as banks lending money, no longer lead to the best allocation of resources or something close: Instead, they lead to the worst allocation of resources! Sound familiar?

So, what to worry about? Don't worry about the end of the universe or the Earth falling into a galactic black hole. But if banks leverage to the hilt again, then you should worry about hearing another big sucking sound.

THE OPINIONS OF SEARCH ENGINES

W. DANIEL HILLIS
Physicist, computer scientist, co-chairman of Applied Minds, LLC; founder, Metaweb Technologies; author, The Pattern on the Stone

Last year, Google made a fundamental change in the way it searches. Previously a search for, say, "Museums of New York" would return Web pages with sequences of letters that matched your search terms, like M-U-S-E-U-M. Now, besides the traditional keyword search, Google also performs a "semantic search," using a database of knowledge about the world. In this case, it will look for entities that it knows to be museums that are located within the geographic region that is named New York. To do this, the computers performing the search must have some notion of what a museum is, what New York is, and how they are related. The computers must represent this knowledge and use it to make a judgment.

The search engine's judgments are based on knowledge of specific entities: places, organizations, songs, products, historical events, even individual people. Sometimes these entities are displayed to the right of the results, which combine the findings from both methods of search. Google currently knows about hundreds of millions of specific entities. For comparison, the largest human-readable reference source, Wikipedia, has fewer than 10 million entries. This is an early example of semantic search. Eventually every major search engine will use similar methods. Semantics will displace the traditional keywords as the primary method of search.

A problem becomes apparent if we change the example from "Museums of New York" to "Provinces of China." Is Taiwan such a province? This is a controversial question. With semantic search, either the computer or the curator of the knowledge will have to make a decision. Editors of published content have long made such judgments; now the search engine makes these judgments in selecting its results. With semantic search, these decisions are based not on statistics but on a model of the world.

What about a search for "Dictators of the World"? Here the results, which include a list of famous dictators, are not just the judgment of whether a particular person is a dictator but also an implied judgment, in the collection of individual examples, of the very concept of "dictator." By building knowledge of concepts like "dictator" into our shared means of discovering information, we are implicitly accepting a set of assumptions.

Search engines have long been judges of what is important; now they are also arbiters of the truth. Different search engines, or different collections of knowledge, may evolve to serve various different constituencies—one for mainland China, another for Taiwan; one for the liberals, another for the conservatives. Or, to put it more optimistically, search engines may evolve new ways to introduce us to unfamiliar points of view, challenging us with new perspectives. Either way, their invisible judgments will frame our awareness.

In the past, meaning was only in the minds of humans. Now it is also in the minds of tools that bring us information. From now on, search engines will have an editorial point of view and search results will reflect that viewpoint. We can no longer ignore the assumptions behind the results.

TECHNOLOGY-GENERATED FASCISM

DAVID BODANIS

Writer; futurist; author, Passionate Minds: Emilie du Châtelet, Voltaire, and the Great Love Affair of the Enlightenment

I'm worried that our technology is helping to bring the long postwar consensus against fascism to an end. Greece was once the cradle of democracy, yet on live television there recently, a leader of the Golden Dawn fascist movement started beating a female MP who disagreed with his views—smashing her on one side of the head, then the other—and his poll ratings went up, not down.

The problem is that our latest technology works best with no one in control. Silicon Valley trumps anything that more regulated economies can produce. But when the results are applied in banking, or management, the consequence for those in the middle and bottom of society often is chaos: jobs coming and going, seemingly at random.

Because of what fascism led to in the past, it's easy to forget how attractive it can be for most citizens in troubled times. With a good enemy to hate, atomized individuals get a warm sense of unity. And although some gentle souls like to imagine, frowningly, that only an ill-educated minority will ever enjoy physical violence, that's not at all the case. Schoolchildren almost everywhere enjoy seeing a weaker child being tormented. Fears about our own weakness disappear when an enemy is mocked and punished—a reflex that radio shock jocks across America skillfully manipulate.

This cry of the dispossessed—this desire for restoring order, this noble punishment of those who "dare" to undermine us—will get a particular boost from medical technology. Medicine is getting better, but it's also getting more expensive. Extrapolate those trends.

There's every reason to think that modifications of Botox will be longer-lasting and avoid unmovable foreheads—but what if they cost $4,000 a shot? The differences in physique noticeable even now between the wealthy and the poor at many shopping malls will only be exacerbated.

Gene therapy is likely to take that further, quite plausibly slowing aging by decades—but what if that costs several hundred thousand dollars? Many people in our cities' wealthiest neighborhoods would start using it. It's not hard to imagine popular leaders, with no memory of World War II, inspiring those outside the select high-income domains, those who are mocked beyond endurance by the existence of these near-immortals in their midst, to pursue this all-too-human response to injustice.

But it's a response that will halt technological innovation in its tracks, at precisely the time when we need it most.

MAGIC

NEIL GERSHENFELD
Physicist; director, MIT's Center for Bits & Atoms; author,
Fab: The Coming Revolution on Your Desktop—from Personal Computers to Personal Fabrication

Arthur C. Clarke famously observed that "any sufficiently advanced technology is indistinguishable from magic." That's what I'm worried about.

2001 has come and long since gone. Once upon a time, we were going to live in a science-fiction future, with flying cars, wrist communicators, and quantum teleporters.

Oh, wait—all of that stuff *is* here. You can today buy cars with wings, smartphone watches, and, for those so inclined, single-photon sources to make entangled pairs. But the future brought the past with it. We can now create life from scratch and model the global climate, yet battles rage over the teaching of evolution or the human impact on the environment—battles that Darwin or Galileo would recognize as challenges to the validity of the scientific method.

There is a cognitive dissonance in the idea of fundamentalists using satellite phones in their quest for a medieval society, or creationists who don't believe in evolution getting flu shots based on genetic analysis of seasonal mutations in influenza virus. These advances are linked by workings that are invisible: Deities behave in mysterious ways, and so do cell phones.

We're in danger of becoming a cargo cult, living with the inventions of ancestors from a mythical time of stable long-term research funding. My word processor is good enough—I don't need radical innovation in text entry to meet my writing needs. The same may be happening to us as a society. If the technologies already available can provide adequate food, shelter, heat, light, and viral videos of cute kittens, invention is no longer the imperative for survival it once was.

The risk in seeing advanced technology as magic is failing to see where it comes from. The ability to distinguish which is which matters for recognizing the difference between progress and nonsense. We do have magic spells—I'm sure that Gandalf would be impressed by our ability to teach sand to play chess (in the form of silicon transistors), or be terrified by our ability to destroy a city with a lump of (uranium) metal. But these incantations came from building predictive models based on experimental observations, not declarations of beliefs. Accepting the benefits of science without having to accept the methods of science offers us the freedom to ignore inconvenient truths about the environment or the economy or education. Conversely, anyone who has done any kind of technical development has had to confront an external reality that doesn't conform to personal interpretation.

Rather than seeking to hide the workings of technology, we should seek every opportunity to expose them. The quest for technologies that work like magic is leading to a perverse kind of technical devolution. Mobile operating systems that forbid users from seeing their own file systems; touch interfaces that eliminate the use of fine motor control; cars that prevent owners from accessing maintenance data—these all make it easier to do easy things but harder to do hard things.

The challenges we face as a planet require finding highest

rather than lowest common denominators. Learning curves that progress from simple to difficult skills should be sought, not avoided. My understanding is that wizards must train for years to master their spells. Any sufficiently advanced magic is indistinguishable from technology.

DATA DISENFRANCHISEMENT

DAVID ROWAN
Editor, WIRED U.K.

In a big-data world, it takes an exponentially rising curve of statistics to bring home just how subjugated we now are to the data crunchers' powers. Each day, according to IBM, we collectively generate 2.5 quintillion bytes—a tsunami of structured and unstructured data that's growing, in the International Data Corporation's reckoning, at 60 percent a year. Walmart drags a million hourly retail transactions into a database that long ago passed 2.5 petabytes; Facebook processes 2.5 billion pieces of content and 500 terabytes of data each day; and Google, whose YouTube division alone gains seventy-two hours of new video every minute, accumulates 24 petabytes of data in a single day. No wonder the rock star of Silicon Valley is no longer the genius software engineer but the analytically inclined, ever-more-venerated data scientist.

Certainly there are vast public benefits in the smart processing of these zetta- and yottabytes of previously unconstrained zeroes and ones. Low-cost genomics allows oncologists to target tumors ever more accurately, using the algorithmic magic of personalized medicine; real-time Bayesian analysis lets counterintelligence forces identify the bad guys, or at least attempt to, in new data-mining approaches to fighting terrorism. And let's not forget the commercial advantages accruing to businesses that turn raw numbers into actionable information: According to the Economist Intelligence Unit, companies that use effective data analytics typically outperform their peers on stock markets by a factor of 250 percent.

Yet as our lives are swept unstoppably into the data-driven world, such benefits are being denied to a fast-emerging data underclass. Any citizen lacking a basic understanding of, and at least minimal access to, the new algorithmic tools will increasingly be disadvantaged in vast areas of economic, political, and social participation. The data-disenfranchised will find it harder to establish personal creditworthiness or political influence; they will be discriminated against by stock markets and social networks. We need to start seeing data literacy as a requisite fundamental skill in a 21st-century democracy, and to campaign—and perhaps even legislate—to protect the interests of those being left behind.

The data-disenfranchised suffer in two main ways. First, they face systemic disadvantages in markets that are nominally open to all. Take stock markets: Any human traders today bold enough to compete against the algorithms of high-frequency and low-latency traders should be made aware of how far the odds are stacked against them. As Andrei Kirilenko, the chief economist at the U.S. Commodity Futures Trading Commission, along with researchers from Princeton and the University of Washington found recently, the most aggressive high-frequency traders tend to make the greatest profits—which suggests that it would be wise for the small investor simply to leave the machines to it. It's no coincidence that power in a swath of other sectors is accruing to those who control the algorithms—whether the Obama campaign's electoral "microtargeters" or the yield-raising strategists of data-fueled precision agriculture.

Second, absolute power is accruing to a small number of data superminers whose influence is matched only by their lack of accountability. Your identity is increasingly what the data oligopolists say it is: Credit agencies, employers, prospective dates, even the U.S. National Security Agency have a fixed view of you based on your online data stream as channeled via search engines, social networks, and "influence" scoring sites, however inaccurate or outdated the results. And good luck trying to correct the errors or false impressions that are damaging your prospects. As disenfranchised users of services such as Instagram and Facebook have increasingly come to realize, it's up to them, not you, as to how your personal data shall be used. The customer may indeed be the product, but there should at least be a duty for such services clearly to inform and educate the customer about his lack of ownership in their digital output.

Data, as we know, is power—and as our personal metrics become ever easier to amass and store, that power needs rebalancing strongly toward us as individuals and citizens. We impeded medical progress by letting pharmaceutical companies selectively, and on occasion misleadingly, control the release of clinical trials data. In the emerging yottabyte age, let's ensure the sovereignty of the people over the databases by holding to account those with the keys to the machine.

BIG EXPERIMENTS WON'T HAPPEN

LISA RANDALL
Theoretical physicist & cosmologist; Frank B. Baird, Jr., Professor of Science, Harvard University; author, Knocking on Heaven's Door: How Physics and Scientific Thinking Illuminate the Universe and the Modern World

I worry that people will gradually stop the major long-term investments in research that are essential if we are to answer difficult (and often abstract) scientific questions. Important fundamental experimental science will always be at the edge of what is technologically feasible, and moving forward requires commitment to advances. The applications are not obvious, so there has to be an underlying belief that finding the answers to deep and significant questions about how the universe evolved, how we evolved, what we are made of, what space is made of, and how things work is important. The ability to find answers to these questions is one of the characteristics that makes human beings unique and gives meaning to our lives. Relinquishing this for short-term ends would be a tragedy.

In my field of particle physics, everyone is worried. I don't say that lightly. Recently I've been to three conferences where the future was a major topic of discussion. Many ideas were presented, but my colleagues and I certainly worry whether experiments will happen.

At the moment, we have the Large Hadron Collider—the giant accelerator near Geneva, where protons collide at very high energies—to turn to for new experimental results. In the summer of 2012, we learned that a Higgs boson exists. It was a major milestone, of which the LHC engineers and experimenters can be proud. With last year's data, where the decays of many more Higgs bosons were recorded, we will understand more about the particle's properties.

But we also want to know what lies beyond the Higgs particle—what is it that explains how the Higgs boson ended up with the mass it has. The LHC also promises to answer this question when it's turned back on in 2015 after having been shut down for two years to upgrade to higher energy.

But the energy gain will be less than a factor of 2. I can fairly confidently say that I expect answers implying the existence of new particles beyond the Higgs boson. But I can't confidently say that I expect them to be less than a factor of 2 heavier than the energies we have already explored. This is worrisome. Not finding anything, ironically, would be the best argument that the LHC energy was simply not high enough and more energy is necessary. But discoveries are what usually egg us on. Not finding anything would be very bad indeed.

The Superconducting Super Collider, which was canceled by the U.S. Congress, would have had about three times the energy. It was designed with the ultimate physics goals in mind, which argued for a more powerful machine. The LHC—though designed to answer similar questions—was built in a pre-existing tunnel, which constrained the maximum energy that could be achieved. If we had three times the energy, I'd be a lot less worried. But we don't.

So I'm worried. I'm worried I won't know the answer to questions I care deeply about. Theoretical research (what I do) can of course be done more cheaply. A pencil and paper, and even a computer, are pretty cheap. But without experiments or the hope of experiments, theoretical science can't truly advance. Happily, advances wouldn't cease altogether, as we would still get new results from astronomical observations and smaller-scale experiments on Earth. And there would be many ideas to play around with. But we wouldn't know which of them represented what is really going on in the world.

On top of that, the universe often has more imagination than we do. We need to know what the universe is telling us. Some of the best new ideas come from trying to explain mysterious phenomena. I hope the future presents us with some answers—but also more mysteries to explain.

THE NIGHTMARE SCENARIO FOR FUNDAMENTAL PHYSICS

PETER WOIT

Mathematical physicist, Columbia University; author, Not Even Wrong: The Failure of String Theory and the Search for Unity in Physical Law

During the 20th century, the search for a theory of how the physical world works at its most fundamental level went from one success to another. The earliest years of the century saw revolutionary new ideas, including Einstein's special relativity and the beginnings of quantum theory, while the decades that followed each of these ideas were times of surprising new insights. By the mid-1970s, all the elements of what is now called the Standard Model were in place, and the century's final decades were dominated by endless experimental results confirming this theory's predictions. By the end of the millennium, we were left in an uncomfortable state: The Standard Model was not fully satisfactory, leaving various important questions unanswered but no experimental results disagreeing with it. Physicists had few hints as to how to proceed.

The Large Hadron Collider was supposed to be the answer to this problem. It could produce Higgs particles, allowing study of a crucial and less than satisfactory part of the Standard Model which had never been tested. A raft of heavily promoted speculative and unconvincing schemes for "Beyond Standard Model" physics all promised exciting new phenomena to be found at LHC-accessible energies.

Results from the LHC have started to come in, and they are carrying disturbing implications. Unsurprisingly, none of the promised "Beyond Standard Model" particles have put in an appearance. More worrisome, though, is the big LHC success: the discovery of the Higgs. Within the still-large experimental uncertainties, now that we've finally seen Higgs particles they look all too much as if they're behaving just as the Standard Model predicted they would. What physicists face now is a possibility they always knew was there but couldn't believe would really come to pass: the "Nightmare Scenario" of the LHC finding a Standard Model Higgs and nothing more.

For the experimentalists, this leaves the way forward unclear. The case for the LHC was obvious: The technology was available, and the Higgs or something else had to be there for it to discover. Going to higher energies is extremely difficult, however, and there's now no good reason to expect to find anything especially new. A lower-energy "Higgs Factory" special-purpose machine designed for detailed study of the Higgs may be the best bet. In the longer term, we may need technological breakthroughs to allow studies of physics at higher energies at affordable cost.

Theorists in principle are immune to the constraints imposed by technology, but they face the challenge of dealing with the unprecedented collapse of decades of speculative work and no help from the experiment on the question of where to turn to for new ideas. The sociological structure of the field is ill equipped to handle this situation. Already we have seen a turn away from confronting difficult problems and toward promoting fatalistic arguments that nothing can be done. Arguments are being made that because of random fluctuations we live in a corner of a "multiverse" of possibilities, with no hope of ever answering some basic questions about why things are the way they are.

These worries are in some sense just those of a narrow group of scientists, but I think they may have much wider implications. After centuries of great progress, moving toward ever deeper understanding of the universe we live in, we may be entering a new kind of era. Will intellectual progress become just a memory, with an important aspect of human civilization increasingly characterized by an unfamiliar and disturbing stasis? This unfortunately seems to be becoming something worth worrying about.

NO SURPRISES FROM THE LHC: NO WORRIES FOR THEORETICAL PHYSICS

AMANDA GEFTER
Consultant, New Scientist; *founding editor, CultureLab*

It's a physicist's worst nightmare: After $9 billion, three decades of calculation, and the dedicated work of 10,000 scientists, CERN's Large Hadron Collider—an underground particle accelerator spanning two nations, the largest machine humankind has ever built, the shining hope of reality hunters worldwide—hasn't turned up a single surprise. No supersymmetric particles, extra dimensions, or unforeseen curiosities have made an appearance. Worse, the Higgs boson that did show up seems to be the very one predicted by the Standard Model. Now the machine has been switched off in preparation for a higher energy run and will remain shut down for nearly two years, leaving theoretical physicists to ponder the cosmic mysteries armed with nothing more than what they already knew.

Physicists are funny that way. Their worst nightmare is to be right. They'd much prefer to be wrong or, better yet, shocked. Being right, they fear, is a dead-end; the lack of surprises at the LHC will stifle progress in theoretical physics. But I see no reason to worry. In fact, if we look back through the history of physics, it's clear that such progress is rarely driven by experimental anomalies.

Einstein thought deeply about the role anomalies play in physics. He distinguished between two types of theory: constructive and principled. Constructive theories, he said, are models cobbled together to account for anomalous observations, like the double-slit experiment that shaped the theory of quantum mechanics. Principled theories, on the other hand, start with logical principles from which particular facts about the world—including anomalies—can be derived. Relativity was a principled theory. "The supreme task of the physicist is to arrive at those universal elementary laws from which the cosmos can be built up by pure deduction," Einstein said. How? By seeking out paradox—that place where established principles collide in contradiction, where logic self-destructs.

The development of the theory of special relativity offers a pristine comparison of the two approaches, since Einstein wasn't the only one to discover its equations. Hendrik Lorentz arrived at them by way of an experimental anomaly—namely, the null results of the Michelson-Morley experiment. Albert Michelson and Edward Morley had set out to measure the Earth's motion relative to the hypothetical ether by splitting a light beam in two and sending each half down the arm of an interferometer, one running parallel to the Earth's motion around the sun, the other perpendicular. At the end of each arm was a mirror, which would bounce the light back to its starting point. Physicists expected that the light beamed down the parallel arm would take longer to return, since it would have to travel against the ether "wind." To everyone's surprise, the beams arrived at their origin at exactly the same time. To account for the anomaly, Lorentz suggested that the ether physically shortened the instrument's parallel arm by the precise amount needed to make up for the missing time delay. How or why this miraculous contraction would occur was a mystery, but Lorentz wrote down the equations to describe the result.

Meanwhile, Einstein arrived at the same equations, only he wasn't looking at the Michelson-Morley anomaly. He had his eye on a paradox instead. He saw that Maxwell's theory of electromagnetism conflicted with the principle that the laws of physics be the same for all observers. Both had to be true, and both couldn't be true. The only way to resolve a paradox is to find and then drop the offending assumption. For Einstein, the culprit was the absolute nature of space and time. In its ruin sat the principle of relativity.

If Lorentz and Einstein both came up with the equations of special relativity, why do we credit the theory to the latter? Because although Lorentz had the right equations, he had no idea what they meant. When Einstein discovered that space and time change relative to an observer's motion so that the speed of light remains the same in every reference frame, he didn't just account for the Michelson-Morley result, he *explained* it.

Paradox has driven our understanding of the universe in recent history as well—specifically, the black-hole information-loss paradox. According to quantum mechanics, information had to escape an evaporating black hole; according to relativity, it couldn't. Both had to be true, and both couldn't be true. This time Leonard Susskind came to the rescue. The offending assumption he fingered was the absolute nature of spacetime locality. In the remains of the paradox, Susskind discovered both the principle of horizon complementarity and the holographic principle.

It's not hard to see that discoveries driven by paradox and principle are more useful than those arrived at through experiment. Understanding nature's principles is just that—understanding. It's what science is all about. Theories kludged together to account for experiment, on the other hand, are ad-hoc by their very nature. They give us the how without the why. Thus quantum mechanics, despite its incredible experimental success, remains a mystery.

That's not to say we don't need experiment. Experiment is the ultimate arbiter of science, tasked with the final say over whether theories are right or wrong. Anomalies are not guides but constraints. Einstein could not have derived general relativity from Mercury's anomalous orbit. But once he discovered the theory through logical principles, its ability to account for the planet's strange perihelion helped convince scientists that the theory was right.

Despite the disappointment echoing through the empty stretches of the LHC, physicists have plenty of anomalies on their hands already. There's the dark energy that's speeding up the universe's expansion, and the unidentified dark matter that tethers stars to their galaxies. There's the lack of large-scale fluctuations in the cosmic microwave background radiation, and the synchronized swim of galaxies dubbed the dark flow. There may even be surprises lurking in the LHC data yet.

But the point is, anomalies don't promise understanding. For that, we need paradox. Luckily one has cropped up just in time. Put forward by Ahmed Almheiri, Donald Marolf, Joseph Polchinski, and James Sully, the "firewall paradox" once again pits general relativity against quantum mechanics in the vicinity of an evaporating black hole, but in this case horizon complementarity doesn't seem to be enough to resolve the paradox.[a] Yet another assumption will have to go. Figuring out which one to abandon will earn physicists profound insights into the workings of the universe. It's a physicists best dream.

CRISIS AT THE FOUNDATIONS OF PHYSICS

STEVE GIDDINGS
Theoretical physicist, UC Santa Barbara

What really keeps me awake at night (besides being sued by an unprincipled water developer in Utah—another story) is that we face a crisis within the deepest foundations of physics. The only way out seems to involve profound revision of fundamental physical principles.

But I'll begin the story with something I believe we shouldn't worry about—"firewalls," a radically deviant picture of black holes recently advocated by colleagues, in which an observer falling into a black hole burns up right at the horizon. The firewall picture—really a clever renaming of a previously known phenomenon—has provoked some recent controversy, which has begun reaching the press. If correct, it could well mean that the black hole at the center of our galaxy is not a black hole at all: Instead, it is the firewall at the center of the galaxy and instantaneously destroys anyone or anything who ventures within about 12 million kilometers of its center.

I'm not worried that this scenario is correct, because it is too preposterous and removed from known physics, and I believe there are less preposterous alternatives. But the fact that otherwise serious physicists have forwarded it is definitely symptomatic of the profound crisis we face.

In short, the crisis is a deep conflict between fundamental physical principles that form the foundation of our most basic framework for describing physics. These pillars are *quantum mechanics*, the principles of *relativity* (special or general), and *locality.* These pillars underlie local quantum field theory, which serves as the basis for our description of physical reality—from the shining sun to the creation of matter in the early universe to the Higgs boson (if that's what it is).

These principles clash when pushed to the extreme; the sharpest version of the problem arises when we collide two particles at sufficient energy to form a black hole. Here we encounter the famed black-hole information problem: If the incoming particles start in a pure quantum state, Hawking's calculation predicts that the black hole evaporates into a mixed, thermal-like final state, with a massive loss of quantum information. This would violate—and thus doom—quantum mechanics. While serious people still consider modifying quantum mechanics, proposals to do so have so far created much bigger problems. For example, it has been argued that Hawking's original proposed modification of quantum mechanics would imply that a "firewall" exists for all observers, everywhere! Quantum mechanics appears to be remarkably resistant to sensible modification.

If quantum mechanics is sacred, apparently other principles must go—either those of relativistic invariance or of locality, or both. The former likewise appears resistant to sensible modifications, but locality is a remarkably "soft" principle, in the context of a theory with quantum mechanics and gravity. So that seems a reasonable concept to suspect.

The basic statement of locality is that quantum information cannot propagate faster than the speed of light. At least as early as 1992, modification of locality to solve the problem of getting missing quantum information out of a black hole was proposed. In the following years, a picture involving an even more radical form of nonlocality took shape. This picture was based on the holographic principle and a new notion of complementarity, proposing that observations inside and outside a black hole are complementary, analogous to Niels Bohr's complementarity of position and momentum measurements in quantum mechanics. Holography/complementarity was widely explored and became accepted by a significant segment of the physics community.

What has become clear in the past few years is that a picture based on complementarity is not only unnecessarily radical but also likely inconsistent. While the jury of the physics community may still be out on complementarity, new awareness of its apparent downfall has rekindled broader interest in the deep crisis we face.

In a context where one or more supposed bedrock principles must be discarded, we obviously need to be a little crazy—but not too crazy! Complementarity is an extreme and plausibly inconsistent form of nonlocality. On the other hand, if quantum mechanics is to be saved by allowing nonlocality inside a black hole—but you accept that quantum field theory holds everywhere outside a black hole—you encounter the firewall problem. Is there a less crazy alternative? I believe so, in the form of a more "nonviolent" nonlocality that extends outside what we would describe as the black hole horizon and transfers quantum information out. I say "would describe" because the horizon concept makes reference to a classical spacetime picture, which may well not be fundamentally correct. More could be said, but describing such nonlocality would violate the editorial marching orders of this forum.

Suffice to say: Whereas it appears that one of three basic pillars of physics must be modified and locality is the prime suspect, modification of locality is no small matter. Naïve modifications of locality—as often proposed by physicists "on the fringe"—generically lead to disastrous collapse of the entire framework of quantum field theory, which not only has been experimentally tested to a very high degree of accuracy but underlies our entire physical picture of the world. If such modification must be made, it must be subtle indeed. It also appears that the basic picture of reality as underlain by the fabric of space and time may well be doomed. What could replace it is a framework wherein the mathematical structure of quantum mechanics comes to the fore. I would say more, but . . . marching orders.

I *will* say that I am deeply concerned about how we will arrive at a complete and consistent theory of gravity. And that we must do so, in order to describe not only black holes (which have been found to be ubiquitous in the universe) but also both the early inflationary and pre-inflationary evolution of our universe, as well as our seemingly dark-energy-dominated future. The current problems at the foundations link to many big questions—and I fear it will be no small feat to resolve them.

THE END OF FUNDAMENTAL SCIENCE?

MARIO LIVIO
Astrophysicist, Space Telescope Science Institute; author,
Brilliant Blunders: From Darwin to Einstein—Colossal Mistakes by Great Scientists That Changed Our Understanding of Life and the Universe

Fundamental physics appears to be entering a new phase. And this new phase has many physicists worried.

Human curiosity about natural phenomena has always exceeded what was merely necessary for survival. On one hand, this curiosity has led to the generation of elaborate mythologies and religions. On the other, it brought about the ascent of science.

The desire to explain the universe and make definitive predictions about cosmic phenomena on all scales has been one of the key drivers of science since the pioneering thinking of Galileo, Descartes, and Newton. The achievements have been truly astounding. In fact, we now have a verifiable "story" for the evolution of the universe from when it was no more than a minute old to the present.

One of the main pillars on which scientists have always constructed their theories (the so-called scientific method) is *falsifiability*—theories have to make clear predictions that can be directly tested by experiments or observations.

The hope has been that we can find a theory that will explain not only all the forces we observe in nature (gravity, electromagnetism, and the two nuclear forces) but also all the values of all the constants of nature (such as the relative strength of the forces, the ratio of the masses of the elementary subatomic particles, and so on). In the past two decades, however, ideas have emerged that maybe some of the constants of nature are not truly "fundamental" but rather "accidental." Instead of one universe, these theories propose, there is really a huge ensemble of universes—a "multiverse."

Those accidental constants take different random values in different members of the ensemble, and the value we happen to observe in our universe is simply one that is consistent with the fact that complexity and life have evolved. Thus there is no true physical explanation for the value of some constants.

The details of these models are unimportant. What is important is that parallel universes may be directly unobservable. And this is what keeps many physicists up at night. A huge number of unobservable universes seems to depart from the scientific method into the realm of pure metaphysics. To some scientists, admitting that some parts of our universe do not have an explanation from first principles marks the end of fundamental science.

Well, does it? I don't think so.

We cannot directly observe free quarks (the constituents of protons), and yet all physicists believe in their existence. Why? Because the theory of quarks (known as the Standard Model) makes enough predictions that *can* be directly verified, so we accept all of its predictions. For the idea of the multiverse and accidental constants to be accepted, it, too, would have to make enough verifiable predictions in the observable universe. If it doesn't, it will remain mere speculation, not a theory.

There is also still a chance that physicists will eventually succeed in formulating a theory that does not require the concept of accidental variables. This will clearly be fantastic, and in the spirit of the historical development of science. But we do have to keep open to the possibility that our dream of a Theory of Everything is based on a misunderstanding of what is truly fundamental. In that case, the multiverse may simply mark the beginning of a new and exciting era of scientific thinking, rather than the end of science.

QUANTUM MECHANICS

LEE SMOLIN
Theoretical physicist, Perimeter Institute; author, Time Reborn

I worry that we don't really understand quantum phenomena. We have a successful theory—quantum mechanics, which has passed every experimental test since it was formulated in 1926 and is the foundation for our understanding of all of physics except gravity. It is the basis of new technologies of quantum information and computation that are currently under intensive development. Still, I don't believe quantum mechanics gives a complete description of nature. I strongly believe there is another, truer description waiting to be discovered.

A large part of the reason is that quantum mechanics gives only probabilistic predictions for the results of experiments. Atoms and molecules come in discrete states, with distinct energies. These discrete states are the basis for our understanding of chemistry. When an atom or molecule transitions between these states, it emits or absorbs photons to make up the jump in energy. Quantum mechanics provides probabilities for these various transitions to occur. The probabilities can be compared with rates measured by experiments done on many systems at a time.

But consider a particular individual atom that decays at a particular instant from one excited state to a lower energy state. Quantum mechanics provides no precise prediction as to when that instant will be, nor does it provide any basis for an explanation of why it happens when it does. I believe there must be a reason why every individual event happens, and that if we understood it we could make a theory to predict in principle when each atom transitions. That is to say, I want a theory of quantum phenomena that goes beyond quantum mechanics.

The people who originally formulated quantum mechanics did not claim that it gave a complete description of physical reality. Bohr, Heisenberg, and their followers explained that there were two kinds of processes in nature—ordinary processes and measurement processes. To distinguish them, they divided the world in two. On one side of the divide is the quantum system we want to describe. On the other side is our world, populated by people, measuring instruments, and clocks. These are to be described by old-fashioned Newtonian physics. A measurement process happens when the two worlds interact and the probabilities describe outcomes of these interactions. Quantum mechanics, they argued, gives no description or account of what goes on in the purely quantum world. It is a tool for discussing the probabilities for different outcomes of interventions we make on a quantum system.

But surely we and our measuring systems are made of atoms and so are part of the quantum world. In that case, shouldn't it be possible to put everything into the quantum side of the division and treat ourselves and our tools as quantum systems? When we try to do this, we get no account of experiments giving definite results and no predictions of probabilities for those results. To make sense of this is called the measurement problem, and it has been under contention for eight decades.

One response is the many-worlds interpretation, which claims that reality is vastly bigger than the world we observe and contains an infinitude of histories, one for every sequence of outcomes for which quantum mechanics predicts probabilities. The world we experience is only one of these histories.

I do not believe in this vast expansion of reality. I believe there is only one real world, in which there are definite outcomes of experiments, in which only one history is realized of the many possible.

Many other resolutions have been offered to accommodate the fact that quantum mechanics does not give an account of individual phenomena. Perhaps logic could be changed so that it becomes no longer possible to express our perplexity. Perhaps quantum mechanics describes not nature but only the information that an observer may gather about a physical system by doing experiments.

But there is another possibility: that quantum mechanics does not provide an explanation for what happens in individual phenomena because it is incomplete—because it simply leaves out aspects of nature needed for a true description. This is what Einstein believed and it is also what de Broglie and Schrödinger, who made key steps formulating the theory, believed. This is what I believe, and my lifelong worry has been how to discover that more complete theory.

A completion of quantum mechanics that allows a full description of individual phenomena is called a hidden-variables theory. Several have been invented; one that has been much studied was invented by de Broglie in 1928 and reinvented by David Bohm in the 1950s. This shows it's possible; now what we need to do is find the right theory. The best way to do that would be to discover a theory that agreed with all past tests of quantum mechanics but disagreed about the outcomes of experiments with large, complex quantum devices now under development.

We know that such a theory must be radically nonlocal, in the sense that once two particles interact and separate, their properties are entangled even if they travel far from each other. This implies that information as to the precise outcomes of experiments they may each be subject to has to be able to travel faster than light.

So a complete theory of quantum phenomena must contain a theory of space and time. I've long believed that the task of completing quantum mechanics and the challenge of unifying quantum mechanics with spacetime are one and the same problem. I also see the problem of extending our understanding of physics at the cosmological scale to be the same as discovering the world behind quantum mechanics.

Many physicists accept quantum mechanics as a final theory and attempt to solve the open problems in physics and cosmology, such as unification and quantum gravity, within the existing framework of quantum physics. I worry that this is wrong and cannot succeed, because quantum mechanics itself must be radically deepened and completed to make further progress in our understanding of nature.

ONE UNIVERSE

LAWRENCE M. KRAUSS

Physicist/cosmologist; director, Origins Project, Arizona State University; author, A Universe from Nothing

You might think that the fact that we live in a relatively quiet suburb of an average galaxy in an apparently uniform universe 14 billion years after the Big Bang would not be large source of worry. However, lately it has been worrying me.

Thanks to the two pillars of 20th-century physics, relativity and quantum mechanics, we now know that the information we can obtain about the universe is limited by our particular circumstances—although these don't really put strict limits on what is ultimately empirically knowable. There may be, however, new limits looming on our ultimate ability to probe nature—limits made manifest by the truly remarkable successes of physical theory and experiment in the past fifty years. Again, the limits are a consequence of our circumstances; at least in principle, they could change the way fundamental science may progress in the future.

In the first place, current physical theories suggest that our universe is probably not unique. Well beyond our ability to directly probe, there may be an infinity of universes, with differing laws of physics and perhaps different characteristics of space and time. This is not necessarily a problem if we're interested in understanding the nature of our particular universe. But it could be that the laws of physics are probabilistic and there's no fundamental reason why they are what they are in our particular universe. Just as an epidemiologist with a single patient can say little about the cause of some condition because it's impossible to know what's normal and what isn't, if we can study only one universe—our own—we may never be able to directly empirically determine whether fundamental laws are indeed fundamental or just accidents. We might be smart enough to derive a theory that explains how the laws of physics are distributed across universes and determines the probability for our particular laws being what they are. But it is equally plausible that without access to a larger sample set, we may never know.

It gets worse. We have discovered that the expansion of our universe is accelerating; the longer we wait, the less we will see of it. Eventually all the galaxies we see now will disappear, and even the evidence of the Big Bang itself will disappear. Although I have argued (unsuccessfully) to Congress that this means we should do cosmology while we still have the chance, we cannot necessarily presume that if we and our intelligent successors keep at it, Nature will reveal more of her secrets.

The very cause of the acceleration may never be possible to pin down. If empty space has energy, then this energy can cause the observed acceleration. But as there is no known laboratory experiment that can probe this energy, our only recourse is to observe the expansion over time. A constant rate of acceleration is consistent with an energy associated with purely empty space, but it is also consistent with a host of other possible sources of trapped energy in some otherwise invisible fields. We may never have a way of knowing. And if this energy is directly associated with the properties of empty space, we may never know why, because that, too, may be an accident, with different energies in different universes. Given just a single universe to probe, we may never know.

As long as we're in the mood for worrying: Even the apparent discovery of the Higgs boson at the Large Hadron Collider (a finding that validates the most remarkable intellectual journey humans have ever embarked on) may yield frustration on the horizon. So far, just the Higgs has been seen. But the Standard Model has a host of weirdnesses, and a plethora of physics ideas have been proposed to explain them. Most of these ideas suggest some new physics that will be seen at the LHC. But what if the Higgs is all that is seen? Then we will have no guidance on where to turn to resolve the underlying puzzles of the Standard Model. If Nature is not kind, the resolution of these problems may well be at a distance and energy scale that is simply unattainable, either because of practical physical constraints or the constraints imposed by small-minded politicians.

Perhaps then, at the extremes of scale, empirical science will reach its limits and we will be reduced to arguing about what is plausible, instead of testing our ideas. I hope this won't be the case. I wouldn't bet on it, either. After all, every time we open a new window on the universe we're surprised, and there are many windows left to open. Whereas worrying helps prepare the mind, I don't believe that any of these possible limits will lead to the end of science itself—or even the end of physics, as naysayers have proposed. There are enough remarkable and perplexing aspects of the universe which we can still measure to keep us going for a long time.

THE DANGEROUS FASCINATION OF IMAGINATION

CARLO ROVELLI

Theoretical physicist, Centre de Physique Theorique de Luminy, Marseille; author, The First Scientist: Anaximander and His Legacy

Worries about imagination might seem odd in a context like *Edge*, where intelligent creativity shines. I myself have often praised visionary imagination and those who can think what nobody else could. But I worry that free imagination is overvalued and carries risks. In theoretical physics, my field, technical journals are replete with multiuniverses, parallel dimensions, scores of particles nobody has ever seen, and so on. Physicists often look down on philosophers, but they are influenced by philosophers' ideas more than they admit. Most have absorbed from Popper, Kuhn, Feyerabend, and other philosophers the idea that science advances by throwing away the past and entertaining novel visions.

The heroes are Copernicus, who dared to send Earth flying around the sun, and Einstein, who dared to imagine space as curved and fused with time. Copernicus and Einstein turned out to be right. Will the current fantasies be equally successful? I feel a strange sense of unease. It is one thing to have ideas; it is another to have good ideas. There is value in producing ideas. There is value in screening them.

A number of my colleagues in theoretical physics have spent their life studying a possible symmetry of nature called supersymmetry. Experiments in laboratories, such as Geneva's CERN, seem now to be pointing more toward the absence than the presence of this symmetry. I have seen lost stares in the eyes of some colleagues: "Could it be?"—how dare nature not conform to our imagination?

The task of separating the good thoughts from the silly ones is hard, of course, but this is where intelligence matters: What should be nurtured? Today many say that, after all, there are no "good ideas" or "bad ideas"—that all ideas can be good. I hear this in philosophy departments from very smart colleagues: "Every idea is right in its own context" or "We don't have to suppress ideas that might turn out to be better tomorrow" or "Everything is better than lack of creativity." We often have to choose between realizing that we do not know or making up pretty stories. To a large extent we live in narrations that we weave ourselves. So why not just go for the sweetest of these? After we have freed ourselves from the close-mindedness of the past, why not feel free? We can create enchanting explanations, images of ourselves, of our great country, of our great society. We can be fascinated by our dreams.

But something tells me we should worry. We live in a real world, where not all stories are equally good, equally effective. One dream out of many is the good one. Few explanations are the correct one. Einstein used to say, "I have no special talents. I am only passionately curious." His curiosity led him to find the diamonds on the dusty shelves, not dreams out in the blue sky. The difficulty is that more often than not, our wild imaginations turn out to write poor drafts with respect to the surprising variety of reality.

Scientific intelligence met the triumphs that have led us here by positing theories and being extraordinarily suspicious about its own products. My worry is that we are going overboard in our contemporary fascination with imagination; in so doing, we risk losing track of the harsh independence of the world from the weakness of our minds.

WHAT—ME WORRY?

J. CRAIG VENTER

Genomicist; cofounder & chairman, Synthetic Genomics, Inc.; founder & chairman, J. Craig Venter Institute; author, A Life Decoded

As a scientist, an optimist, an atheist, and an alpha male, I don't worry. As a scientist I explore and seek understanding of the world(s) around me and in me. As an optimist I wake up each morning with a new start on all my endeavors, with hope and excitement. As an atheist I know I have only the time between my birth and my death to accomplish something meaningful. As an alpha male I believe I can, and I do, work to solve problems and change the world.

There are many problems confronting humanity, including supplying enough food, water, housing, medicine, and fuel for our ever expanding population. I firmly believe that only science can provide solutions for these challenges, but the adoption of these ideas will depend on the will of governments and individuals.

I'm somewhat of a Libertarian, in that I don't want or need the government to dictate what I can or cannot do to guarantee my safety. For example, I ride motorcycles, sometimes at high speeds; I have full medical coverage and should not be required by the government to wear a helmet to avoid harming myself if I crash. I do wear a helmet, and full safety gear, because I choose to protect myself. Smoking is in a different category. Smoking is clearly harmful to health, and the best step smokers can take to change their medical outcomes is to quit smoking. If that were all there was to it, then the government shouldn't regulate smoking unless it's paying for the smokers' health care. However, science has shown that secondhand smoke can have negative health consequences on individuals in the vicinity of a smoker. Therefore laws and rules to regulate where people can smoke are, in my view, not only reasonable but good for society as a whole.

It's the same with vaccinations. One of the consequences of our ever expanding global population—particularly when coupled with poor public health, unclean water, and misuse of antibiotics—has been and will continue to be new emerging infections, including those from zoonotic outbreaks. Over the past several decades, we have seen the emergence of AIDS, SARS, West Nile, new flu strains, and methicillin-resistant *Staphylococcus aureus* (MRSA). In 2007, MRSA deaths in the United States surpassed HIV deaths. Infectious disease is now the second cause of death in the world, right behind heart disease and ahead of cancer. In 2011, in the U.S., there were twice as many deaths from antibiotic resistance as from automobile accidents.

There are many causes for the emergence of infectious diseases, but one significant factor is human behavior with regard to immunizations. The supposed link between immunizations and autism, which has been proved false, has led some parents to choose not to vaccinate their children, believing it to be a civil liberty issue akin to the choice of whether or not to wear a motorcycle helmet. But I contend that people who avoid immunizations are a major contributing factor to the reemergence and spread of infectious disease, in a way far more dangerous than secondhand smoke. Vaccines are the most effective means of preventing the spread of infectious diseases. There are no better examples than the elimination of polio and smallpox through mandatory vaccinations.

When new or old infectious agents such as viruses and bacteria infect the nonimmunized, genetic recombination can occur, creating new versions of such agents, which can then infect the population that was immunized against the existing strains. We see this occurring with almost every type of infectious pathogen, and—most troubling—we are seeing it here in our own industrialized, wealthy, educated country. There are pockets of outbreaks of diseases such as whooping cough; the emergence in the Middle East of a novel disease-causing coronavirus; illness at Yosemite National Park caused by hantavirus; and the emergence in farm communities of a variant influenza virus (H3N2v) that spread from swine to people. Last year's flu came earlier and was more virulent than in previous years; Boston declared a state of medical emergency because of the number of flu cases and deaths.

Avoidance of vaccination creates a public health hazard. It is not a civil liberty issue. The unvaccinated, coupled with antibiotic resistance and with the decrease in animal habitats (which promotes zoonotic transfer of disease-causing agents) together make for a potential disaster that could take humanity back to the pre-antibiotic era. I thought we learned these lessons after such global pandemics as the plague and the outbreak of 1918 flu that killed 3 percent of the world's population, but clearly without modern science and medicine we will be destined to relive history.

OUR INCREASED MEDICAL KNOW-HOW

ESTHER DYSON

Board member & investor in health care, human capital, & aerospace startups; chairman, EDventure Holdings; former chairman, ICANN (Internet Corporation for Assigned Names & Numbers); author, Release 2.1: A Design for Living in the Digital Age

We should be worried about the consequences of our increasing knowledge of what causes disease and how that knowledge will affect human freedom.

It's exciting that we can figure out what kind of diet and behavior will keep most people well, and it's good to use that knowledge. But that raises the question of who bears the responsibility, if people don't. Indeed, does society have the right to interfere beforehand precisely because society bears some of that responsibility, if only by assuming the burden of health-care costs? It calls to mind the medical irony of the cure being worse than the disease or the immune reaction worse than the pathogen.

In this case, the trigger is our increasing knowledge of how we make ourselves sick—or how we can keep ourselves healthy: proper diet, regular exercise, no smoking, limited drinking, sufficient sleep. It's clear that a squeaky-clean lifestyle (and just for good measure, add avoidance of stress) is the key to good health for most people and to reducing medical costs for society as a whole.

In addition, our knowledge of genetics and specific markers for susceptibility to disease is growing. Certain people with genetic predispositions to certain conditions need to take additional measures to stay healthy. Over time, we will know more of these specific correlations and be able to identify—if we want—people with predisposition to certain diseases (and therefore increased responsibilities?).

The thing to worry about is how society handles this knowledge, both in general and for the specific people who carry extra burdens. We all know of artists and others who are at least somewhat crazy. Many of them avoid treatment for fear—justified or not—of losing their creative gifts. Other people simply want to be themselves rather than some medicated version. And while much medical knowledge may be true, not all of it is. Certainly it keeps changing. Many drugs and other treatments don't do much; others cause collateral damage. How much is an extra month of life worth, if you're diminished by the side effects of life prolongation?

In short, the notion of unpredictable health catastrophes is giving way to something closer to flood plains. People are rightly asking whether society should pay to protect people who live in predictably dangerous areas, refuse to wear helmets while riding motorcycles, and so on. But when somebody actually undergoes such a catastrophe, even a predictable one, attitudes change and the government offers disaster relief. Such rare occasions are becoming more common.

So the questions are: What duty do we have to live *properly?* What responsibility do we have for the consequences if we do not? How much can we blame on our parents, or society, or whomever—and even so, what responsibilty do we bear? Should society pay for prevention but not for remediation of avoidable outcomes? Should we force special responsibilities on people with particular vulnerabilities?

These questions aren't new, but they will become more urgent as we get better at predicting or avoiding outcomes. There are no simple answers to define or to allocate responsibility. That's why we should be worried.

THE PROMISE OF CATHARSIS

ANDRIAN KREYE
Editor, The Feuilleton *(arts and essays), of the German daily* Süddeutsche Zeitung, *Munich*

There are no cathartic moments in evolution. There are plenty in faith and ideology. With the former being a mostly private affair in modern society and the latter a remnant of the 20th century, one should not have to worry about the lure of catharsis anymore. Still there it is, keeping those doors to irrational groupthink open by a tiny crack.

It might not be possible ever to get rid of this powerful lure, though; in the Western world this mindset has been embedded in the cultural DNA too deeply. There is no art form lacking built-in mechanisms that simulate catharsis. Most have been derived from religion.

Take the best-known and most powerful of examples—Beethoven's Ninth Symphony. There are good reasons why this work is performed whenever there is the need to soothe a grieving collective or enhance a moment of national unity. After the attacks of September 11, there were countless performances of the piece worldwide. The divided, cold-war Germany played it whenever sports teams from the two sides joined forces. The European Union has chosen the "Ode to Joy" from the fourth movement as its anthem. It's easy to hear how this works. The key of the Ninth is the tragic D-minor. Over the course of an hour, this somber mood is lifted slowly, until the fourth movement breaks into the jubilant chorus based on Friedrich Schiller's poem. This is solid 19th-century instant catharsis. And there's no doubt where this joy comes from. The poem begins, "Joy, beautiful spark of Gods."

Just as Beethoven's Ninth resorts to the dynamics of a church service, with a sermon promising a rapturous end to all worries, pop culture riffs on the spiritual lift. Take a great rock or soul song or concert. In the perfect case, there will be an exhilarating opening, after which the rhythm will slow down below the speed of a heartbeat. If the drumming and the groove are convincing, the human pulse will adapt. Step by step, the music will accelerate and the rhythms speed up past the regular heartbeat. Lighting, movement, and volume will help to create a state of ecstasy. Those tricks are borrowed from faiths like Voodoo or Pentecostalism. Ever wonder why U2 concerts are always experienced as such an ecstatic event? They openly borrow from the traditions of Catholicism.

The same leitmotifs of catharsis can be found in literature, theater, and film. The classic three-act drama still taught in film school is built like a holy book: setup, conflict, resolution. In narrative art, catharsis isn't simply a reference. When Aristotle came up with the structure of tragedy in the 4th century B.C., the catharsis of the audience's emotions was his avowed goal.

The urge to experience catharsis is, of course, so strong because it always embodies salvation. After the Rapture, there is Paradise. This dynamic has even transcended to the scientific. What else would the Singularity—the belief that artificial intelligence will at some point surpass human intelligence—be, but a technological Rapture absolving humanity from being the ultimate responsible party on this planet?

The problem with catharsis, though, is that it will always remain an empty promise. There is no paradise, no salvation, no ultimate victory. Progress, be it biological, scientific, or social, is a tedious process of trial and error. If we work toward an unobtainable goal, much effort is wasted, and the appearance of false prophets is almost a given. Catharsis thus becomes the ultimate antagonist of rational thinking. If there is Paradise in the Beyond, why bother with the Here-and-Now?

But there's no way to change our mental blueprint. The simulation of catharsis is the very way in which we enjoy art, music, stories, even sports. Every joke's punch line, every hook of a song, promises this tiny moment of release. That is all there is, after catharsis. To ask us to refrain from giving in to this urge would lead to joyless forms of Puritanism. If we are aware of the patterns and dynamic of catharsis, though, it is possible to see the thresholds of escapism.

I'VE GIVEN UP WORRYING

TERRY GILLIAM
Screenwriter, animator, actor; member, Monty Python; director, Brazil, Fear and Loathing in Las Vegas

I've given up worrying. I merely float on a tsunami of acceptance of anything life throws at me . . . and marvel stupidly.

OUR BLIND SPOTS

DANIEL GOLEMAN
Psychologist; author, Leadership: The Power of Emotional Intelligence

Human systems of transportation and energy, construction, and commerce, a scientific consensus tells us, are degrading the global systems that support life on our planet. This damage poses an enormous long-term threat to life as we know it. And yet these changes are too macro or micro to be noticed directly. The amygdala tunes out.

That indifference to danger bespeaks a blind spot built into the brain. Our neural system for threat was attuned to dangers of the Pleistocene: the rustling in the thicket that might signal a predator lurking. But we have no perceptual apparatus or circuitry for alarm that tunes us to the dangers we now face as a species. The brain's perceptual misalignment in recognizing dangers has reached a historic danger point. The human gene pool was robust enough to survive the failure of people in the Middle Ages to see the dangers in rats and fleas as vectors for the Black Plague. But today our collective blind spot has set in motion long-term dynamics in geophysical systems, such as the carbon cycle, that will have devastating impacts not just on humans but on the survival of countless other species—and will take hundreds of years to reverse.

Much of the ongoing damage stems from systemic side effects of our industrial platforms. For instance, we make concrete, bricks, glass, and steel by heating ingredients at very high temperatures for long periods—a technology with roots in the Bronze Age. When these methods were developed, no one saw the coming of the current Anthropocene Age, where human activity pushes systems like the nitrogen and carbon cycles toward the brink.

Now that we can conceptualize the dangers we face, ideally we would put on the brakes. The failure of global meetings at Kyoto and Copenhagen to come to international agreements on slowing global warming is but one symptom of our blind spot at play. But a purely intellectual understanding of a danger does not mobilize our motivational systems all that well.

I happened to attend a closed-door meeting of the heads of sustainability for more than two dozen global companies. Each made a report of what they were doing in their supply chains to combat global warming—a surprisingly encouraging list. But all were agreed that there was a common obstacle to pushing much further: Customers do not care.

To be sure, there are more and more signs that clever ways around our blind spot can be found. For instance, an industrial ecologist at the Harvard School of Public Health has developed "handprints." Instead of tracking all the bad news about our carbon footprint, Handprinter puts a positive spin on our environmental impacts, tracking all the *good* we do. The program lets you measure every action you take that lowers your carbon footprint, and encourages you to keep growing that number.

A proposed Handprinter mobile app would allow you "to interpret your positive impacts in terms of the number of hours or days in which you are living net-positively; use your phone to see the actions other users are taking; you and your friends will be able to keep track of the impact you're having as a group. What's the total collective handprint of your family, neighborhood or school."

Not so long ago, I browsed an upscale magazine that touts life's luxuries. The Christmas issue featured twenty-one "ultimate gifts," including a personally designed fountain pen for $275,000 and a custom-built backyard theme park, including a full-size steel roller coaster, for $20 million plus upkeep. The same week saw the failure of Handprinter to raise funds for that mobile app on the Web site Kickstarter. Price: $30,000.

Albert Einstein once said, "With the unleashed power of the atom everything has changed, save our modes of thinking, and thus we drift towards unparalleled catastrophe." Effective worrying focuses our attention on a genuine threat and leads to anticipating solutions and implementing them. Neurotic worrying just loops into repetitive angst. I'll dial down my worry the day Handprinter not only gets funded but goes viral—and that backyard theme park gets recycled.

THE ANTHROPOCEBO EFFECT

JENNIFER JACQUET

Clinical assistant professor of environmental studies, NYU; researcher, cooperation & the tragedy of the commons

Humans are today something we have never been before, a global geological force. This epoch, which begins around 1800, has been called the Anthropocene and is characterized by steep line graphs that look like Mount Everest sliced in half: human population, water use, biodiversity loss, nitrogen runoff, atmospheric carbon dioxide, and so forth.

The data irrefutably establish humans as the dominant driver of environmental change, which is something that should worry us all. But we should also be worried that framing humans as the dominant driver of change will lead to further negative change, which I am calling *the anthropocebo effect.*

The effects of cultural framing are everywhere. The placebo effect—experiencing positive effects from an inert pill—occurs only in cultures that believe taking a pill can cure an illness. The even stranger nocebo effect, whereby just mentioning side effects makes them more likely to occur, shows us the power of mind. The anthropocebo effect, then, is a psychological condition that exacerbates human-induced damage—a certain pessimism that makes us accept human destruction as inevitable.

Science helps shape how we see ourselves. Words also matter for perception, and perception matters for behavior. Consider what the theory of natural selection did to our view of humans in the biological world. We should be worried about the new epoch, the Anthropocene—not only as a geological phenomenon but also as a cultural frame.

THE RELATIVE OBSCURITY OF THE WRITINGS OF ÉDOUARD GLISSANT

HANS ULRICH OBRIST
Curator, Serpentine Gallery, London; editor, A Brief History of Curating*; coauthor (with Rem Koolhaas),* Project Japan: Metabolism Talks

Although he is not widely known, Édouard Glissant (1928-2011) is one of the most important 20th-century writers whose thinking remains fundamental for the 21st century. I believe we should be worried about his relative obscurity, since he discusses with great insight what seem to me to be the most important issues surrounding globalization: homogenization and extinction. His theory of the "creolization of the world" pertains to questions of national identity in view of the colonial past that characterizes his Antillean identity. He broaches the urgent questions of how best to escape the threat of cultural homogenization and how we can work to sustain the positive force of creolization—a plurality of cultures—within the terms of an ongoing global exchange.

Global homogenization is a tendency that Stefan Zweig had already observed in 1925, when he wrote, in *Die Monotonisierung der Welt* (*The Monotonization of the World*):

> Everything concerning the outer life-form becomes homogeneous; everything is evened to a consistent cultural scheme. More and more countries seem to be congruent, people acting and living within one scheme, more and

more cities becoming similar in their outer appearance. . . . More and more the aroma of the specific seems to evaporate.

The forces of globalization are affecting the world of art at large and exhibition-curating specifically. There has been great potential in the new global dialogs of the last couple of decades, some of it realized, but there has also been the persistent danger that the homogenizing influences of globalization will make differences disappear. I worry about this, and so I read Glissant every morning when I wake up. He anchors my thoughts regarding producing shows internationally—encouraging me to listen to and learn from whatever culture I may be working within. Since time is losing its local variety to a global speed that leaves no room for individual pace, to curate time, and to locate forms of resistance to the homogenization of time, has become as important as curating and resisting the homogenization of space.

Cultural homogenization is nothing less than cultural extinction. As the art historian Horst Bredekamp writes in his book *Theorie des Bildakts* (*Theory of Picture Acts*), in today's globalized war, iconoclastic acts have become a prominent strategy—the public extinction of monuments and cultural symbols, such as the destruction of the Bamiyan Buddha statues in Afghanistan by the Taliban. Through the mass media, war thus becomes a globalized image-war that crosses the borders of territories and persists after the event, functioning as legitimation for military action and diplomatic policy.

Scientists increasingly debate the possibility of the extinction of human civilization and even of the species itself. The astronomer Martin Rees talks about "Our Final Hour" and questions whether civilization will survive beyond the next century. The spectre of extinction is felt across the humanities, too. For philosopher Ray Brassier, the inevitable fact of our eventual extinction grounds the ultimate meaninglessness of human existence, and thus for him the only proper response for philosophy is to embrace and pursue the radically nihilistic implications of this most basic insight. As he writes in *Nihil Unbound* (2007): "[N]ihilism is . . . the unavoidable corollary of the realist conviction that there is a mind-independent reality which . . . is indifferent to our existence and oblivious to the values and meanings which we would drape over it in order to make it more hospitable."

Most would stop short of this absolute nihilism, and there are of course other places to look—sources of hope and meaning. The artist Gustav Metzger, for instance, has made extinction a central theme of his practice. In works that make use of his enormous archive of newspapers, he stresses the point that the prospect of human extinction is continually raised in the innumerable little extinctions that occur in the world. By representing newspaper stories on the subject, Metzger highlights the problem of our collective attitude of resignation in the face of the sheer regularity of this disappearance and our apparent powerlessness to halt it.

Marguerite Humeau is a young artist whose work reaches back before the dawn of human history itself. In what the historian Eric Hobsbawm might have called "a protest against forgetting," she is engaged in a project to reconstruct the vocal cords of extinct animals from prehistory—mammoths, dinosaurs, and others—as a way to bring the sounds of the past back to life. Like Metzger's works, her acts of remembrance contain a warning of a future potentially lying in wait. Humeau's work also makes plain the complex technology enabling her research, a strategy pointing to our collective dilemma: that it has required immense scientific and technological advances for us to become aware of our plight, advances that are often also at the heart of the problem.

Gerhard Richter says that art is the highest form of hope. I would add that art is the primary form of resistance to homogenization and extinction. To quote Zweig again: "Art still exists to give shape to multiple ways of being."

THE DANGER OF INADVERTENTLY PRAISING ZYGOMATIC ARCHES

ROBERT SAPOLSKY
Neuroscientist, Stanford University; author, Monkeyluv: And Other Essays on Our Lives as Animals

I don't think there is free will. That conclusion first hit me in some sort of primordial ooze of insight when I was about thirteen years old, and it has only become stronger since then. What worries me is that although I think this without hesitation, there are times when it's simply too hard to *feel* as though there's no free will—to believe it, to act accordingly. What really worries me is that it's so hard for virtually anyone to truly act as if there's no free will—and that this can have some pretty bad consequences.

If you're a neuroscientist, you might be able think there's free will if you spend your time solely thinking about, say, the kinetics of one enzyme in the brain, or the structure of an ion channel, or how some molecule is transported down an axon. But if you devote your time to thinking about what the brain, hormones, genes, evolution, childhood, fetal environment, and so on have to do with behavior, as I do, it seems simply impossible to think there is free will.

The evidence is broad and varied. Raising the levels of testosterone in someone makes him more likely to interpret an emotionally ambiguous face as a threatening one (and perhaps act accordingly). A mutation in a particular gene increases the odds that she will be sexually disinhibited in middle age. Spending fetal life in a particularly stressful prenatal environment increases the likelihood of overeating as an adult. Transiently inactivating a region of the frontal cortex will render someone more cold-hearted and utilitarian when making decisions in an economics game. Being a psychiatrically healthy first-order relative of a schizophrenic increases the odds of believing in "metamagical" things like UFOs, extrasensory perception, or literalist interpretations of the Bible. Having a normal variant of the gene for the vasopressin receptor makes a guy more likely to have stable romantic relationships. The list goes on and on (and just to make a point that should be obvious from this paragraph but which can't be emphasized too frequently, lack of free will doesn't remotely equal anything about genetic determinism).

The free-will concept requires us to subscribe to the idea that despite the swirl of biological yuck and squishy brain parts filled with genes and hormones and neurotransmitters, there's an underground bunker in a secluded corner of the brain—a command center containing a little homunculus who chooses your behavior. In that view, the homunculus might be made of nanochips, or ancient, dusty vacuum tubes, or old crinkly parchment, or stalactites of your mother's admonishing voice, or streaks of brimstone, or rivets made out of gumption. In that view, whatever the homunculus is made of, it ain't made of something biological. But there *is* no homunculus, and no free will.

This is the only conclusion I can reach. Still, it's hard to believe it, to feel it. I'm willing to admit that I have acted egregiously at times as a result of that limitation. My wife and I get together for brunch with a friend who serves fruit salad. We exclaim, "Wow, the pineapple is delicious!" "They're out of season," our host smugly responds, "but I lucked out and was able to find a couple of good ones." And in response to this, the faces of my wife and I communicate awestruck worship—you really know how to pick fruit, you are a better person than we are. We are praising the host for this display of free will, for the choice made at the split in the road that is Pineapple Choosing. But we're wrong. Genes have something to do with our host's olfactory receptors, which help him to detect ripeness. Maybe he comes from a people whose deep and ancient cultural values include learning how to feel up a pineapple to tell whether or not it's good. The sheer luck of the socioeconomic trajectory of our host's life has provided him with the resources to prowl an overpriced organic market that plays Peruvian folk Muzak.

It's hard to feel as though there's no free will—to not fall for the falsehood that whereas there's a biological substrate of potentials and constraints, there's a homunculus-like separation in what the person has done with that substrate. ("Well, it's not her fault if nature has given her a face that isn't the loveliest, but after all, whose brain is it that chose to get that hideous nose ring?")

This issue transcends mere talk of nose rings and pineapples. As a father, I am immersed in the community of neurotic parents frantically trying to point our children in the direction of the most perfect adulthoods imaginable. When considering our kids' schooling, we cite a body of wonderful research by Carol Dweck, a colleague of mine. To wildly summarize and simplify it: Take a child who has just done something laudable academically and, indeed, laud her—"Wow, that's great, you must be so smart!" Alternatively, in the same circumstance, praise her instead with "Wow, that's great, you must have worked so hard!" The latter is a better route for improving academic performance: Don't praise the child's intrinsic intellectual gifts, praise the effort and discipline she chose to put into the task.

Well, what's wrong with that? Nothing, if the research simply produces value-free prescription. But it is wrong if you are patting the homunculus on the head, concluding that a child who has achieved something through effort is a better, more praise-worthy producer than a child running on plain raw smarts. That's because free will falls by the wayside even when considering self-discipline, executive function, emotional regulation, and gratification postponement. For example, damage to the frontal cortex, the brain region most intimately involved in those functions, produces someone who knows the difference between right and wrong yet still can't control his behavior, even his murderous behavior. Different versions of a subtype of dopamine receptor influence how risk-taking and sensation-seeking a person is. Someone infected with the protozoan parasite *Toxoplasma gondii* is likely to become subtly more impulsive. There's a class of stress hormones that can atrophy neurons in the frontal cortex; by early elementary school, a child raised amid the duress of poverty tends to lag behind in the frontal cortex's maturation.

Maybe we can come to fully realize that when we say, "What beautiful cheekbones you have," we're congratulating the person based on the unstated belief that she chose the shape of her zygomatic arches. It's not that big a problem if we can't achieve that mindset. But it *is* a big problem if, when addressing, say, a six-year-old whose frontocortical development has been hammered by early life stress, we mistake his crummy impulse control for lack of some moral virtue. Or to do the same in any other realm of the foibles and failures, even the monstrosities, of human behavior. This is extremely relevant for the criminal justice system. And to anyone who says that it's dehumanizing to claim that criminal behavior is the end product of a broken biological machine, the answer must be that it's a hell of a lot better than damning the behavior as the end product of a rotten soul. Likewise, it's inadvisable to think in terms of praise, good character, or good choice when looking at the end products of lucky, salutary biology.

But it's difficult to believe there's no free will when so many of the threads of causality are not yet known—or are as intellectually inaccessible as having to weigh the behavioral consequences of everything from the selective pressures of hominid evolution to what someone had for breakfast. This difficulty is something we should all worry about.

THE BELIEF OR LACK OF BELIEF IN FREE WILL IS NOT A SCIENTIFIC MATTER

HOWARD GARDNER
Developmental psychologist; Hobbs Professor of Cognition & Education, Harvard Graduate School of Education; author, Truth, Beauty, and Goodness Reframed: Educating for the Virtues in the Age of Truthiness and Twitter

Over more drinks and more meals than I care to remember, I have argued with colleagues and friends about the existence of free will.

Drawing on recent findings in neuroscience about delayed conscious awareness of actions and reactions but also on a determinist view of causality, most of my conversationalists have insisted that there is no such thing as free will.

For my part, looking at events in ancient and recent history and reflecting on the sometimes surprising conscious life decisions made by others and myself (Martin Luther!), I argue with equal vigor that human beings have free will and that exercising it is what distinguishes us from other animals.

Of course there are various compromises possible: Daniel Dennett–type views that we should act "as if" there were free will and treat others the same way; William James's "My first act of free will shall be to believe in free will"; or Daniel Kahneman's implication that System 1 is automatic, whereas System 2 involves conscious reflecting.

But I've come to the conclusion that the belief or lack of belief in free will is not a scientific matter. Rather, to use the term created by the historian of science Gerald Holton, we are dealing here with themata—fundamental assumptions that scientists and other scholars bring to the work they do. And as such, the existence or the denial of free will, like the question of whether human nature is basically universal or inherently varied, is not one that will ever be settled. And so I have stopped worrying about it.

NATURAL DEATH

ANTONY GARRETT LISI
Theoretical physicist

Within 100 years, unless we make some major research breakthroughs, you are going to die. Before then, you'll get to watch many of the friends and family you love go through the process of decay, infirmity, and death—their personalities and all their memories gone forever. Witnessing this process over and over, we know it is nature's way, but it is horrible, tragic, and heartbreaking. The prospect of our own personal death is so terrifying that most people aren't willing to confront it honestly. To stare into the void and know—truly know—that our existence will shortly end: As natural as it is, we are not emotionally or intellectually equipped to accept it. Instead, most people lie about it to themselves and others.

The greatest lie ever told was that there is a mystical afterlife. This lie has been used for millennia to steel the courage of young men before sending them to kill and die in wars. Even worse, most people lie to themselves when confronting suffering and loss, with stories of a better life after this one, despite there being no credible evidence for any such thing.

But why is it so damaging to share and believe pleasant fantasies of an afterlife, when nonexistence is both inevitable and too horrific to confront? It is damaging because it leads to bad decisions in this life, the only one we have. Knowing that our lives are so short makes each moment and each interaction more precious. The happiness and love we find and make in life are all we get. The fact that there is no supernatural being in the universe that cares about us makes it that much more important that we care about one another.

Because most people lie to themselves about death instead of worrying about it, we are neglecting a radical possibility brought about by advancing technology: Our imminent nonexistence might not be as inevitable as we think. There is no reason, from physics, why our healthy human lives cannot be radically extended by thousands of years or longer. It's just a tricky engineering problem. But for various psychological, social, and political reasons, we're not working very hard on it.

There are, of course, some arguments against extending life. The most common argument is that old age and death are natural. But all diseases cured by technology, including polio, smallpox, and leprosy, were natural but clearly not desirable. Nature can be beautiful but also horrible, and we have been able to alleviate some of that horror through science and technology. The second most common argument is that dramatically extended human life spans would put an unsustainable strain on the environment.

But a person becoming immortal would put no greater strain on the environment than having a child. Also, a rejuvenated person requires less expensive medical treatment than a person disabled by the decay of aging and is more motivated to take care of the future environment, because the chances are greater that he or she will be living in it. A third argument against curing aging is that all attempts at significantly extending human life span have failed. This is true, but this was the same argument holding that the sound barrier couldn't be broken, before Chuck Yeager went rocketing through it.

Even if the probability of quickly finding a technological method to delay or reverse senescence is low, we have been devoting far too little effort to it. After all, no matter what else we might achieve with our work in life, we soon won't be around to enjoy it. There are other problems on the planet to worry about, but none more personally important. And yet, despite this motivation, there is very little money being spent on longevity research. Because there is no history of success, and because of widely held religious beliefs, government won't fund it. And because achieving success will be difficult, and the marketplace is flooded with false claims, industry has little interest in solving the problem. Although the profit could be astronomical, there is no easy path to attain it, unlike the prospects for cosmetic improvements. Over 100 times more money is spent on R&D for curing baldness than for curing aging. We may someday find ourselves with extended life spans as an unintended side effect of taking a pill that gives us fuller hair.

This absurd situation is typical for high-risk, high-reward research in an area without an established record of success. Even with strong motivation, financial support is nearly nonexistent. Scientists working on life extension often lack for equipment or a livable salary and risk their careers by conducting oddball research that repeatedly fails. The problems are hard. But even with limited resources, a handful of scientists are devoting their lives to the pursuit because of what's at stake. Success will require research on a scale similar to that of the Manhattan Project, but government and industry won't be supporting it. The greatest hope is that private individuals will step forward and fund the research directly or through organizations established for that purpose. Maybe an eccentric, farsighted billionaire will want a chance at not dying. Or maybe many people will contribute small amounts to make it happen. This is being done, to some extent, and it gives me hope.

Personally, I know I am not so different from other people; I also have a very difficult time accepting mortality. When I think about all who have been and will be lost, and of my own impending nonexistence, it makes me ill. It's entirely possible that the hope I have for a technological solution to aging and death is biased by my aversion to the abyss. Realistically speaking, although I'm hopeful that radical life extension will happen before I die, given our current rate of technological advance it's more likely that I'll just miss it. Either way, whether aging is cured within my lifetime or afterward, it won't happen soon enough. Good people are suffering and dying, and that needs to change in a way that's never been done before.

THE LOSS OF DEATH

KATE JEFFERY

Professor of behavioral neuroscience & head, Research Department of Cognitive, Perceptual, & Brain Sciences, University College London

In every generation, life distills the best of itself, packages it up and passes it on, shedding the dross and creating a fresher, newer, shinier generation. It has been doing this here for almost 4 billion years and in so doing has transmogrified from unicellular microorganisms that did little more than cling to rocks and photosynthesize to creatures of boundless energy and imagination who write poetry, make music, love one another, and work hard to decipher the secrets of themselves and their universe.

And then they die.

Death is what makes this cyclical renewal and steady advance in organisms possible. Aging and death permit a species to grow and flourish. Because natural selection generally ensures that the child-who-survives-to-reproduce is an improvement on the parent (albeit infinitesimally so, for that is how evolution works), it is better for the parent to step out of the way and allow its (superior) child to succeed in its place. Put more simply, death stops a parent from competing with its children and grandchildren for the same limited resources. So important is death that we have, wired into our genes, a self-destruct senescence program that shuts down operations once we have successfully reproduced, so that we eventually die, leaving our children—the fresher, newer, shinier versions of ourselves—to carry on with the best of what we have given them: the best genes, the best art, and the best ideas. Four billion years of death have served us well.

Now all this may be coming to an end, for one of the things we humans, with our evolved intelligence, are working hard at is trying to eradicate death. This is an understandable enterprise, for nobody wants to die—genes for wanting to die rarely last long in a species. For millennia, human thinkers have dreamed of conquering old age and death. The fight against it permeates our art and culture and much of our science. We personify death as a spectre and loathe it, fear it, and associate it with all that is bad in the world. If we could conquer it, how much better life would become.

Over the last century, that millennia-old dream began to take form, for we humans discovered genes, and within the genes we discovered that there are mechanisms for regulating aging and death, and also that we can engineer these genes—make them do things differently. We can add them, subtract them, alter their function, swap them between species—the possibilities are exciting and boundless. Having discovered the molecular mechanisms that regulate senescence and life span, we have begun to contemplate the possibility that we can alter the life course itself. We may be able to extend life, and possibly quite soon. It has recently been estimated that due to medical and technical advances, the first person to reach the age of a hundred and fifty has already been born. Once we have eradicated cancer, heart disease, and dementia, our biggest killers, we can turn next to the body clock—the mechanism for winding-up operations that limits our life spans—and alter that too. Why stop at a hundred and fifty? If a person is kept disease-free and the aging clock is halted, why could a person not reach the age of two hundred? Three hundred? Five hundred?

What a wonderful idea! Few people seem to doubt that this *is* a wonderful idea, and so research into aging and life span is a funding priority in every wealthy, technologically advanced society. Termed "healthy aging," this research really means prolonging life, for aging is by definition progressive time-dependent loss of health and function, and if we prevent that, we prevent death itself. Who wouldn't want to live to five hundred? To live a life free of decrepitude and pain, to be able to spend much more time enjoying favorite activities, achieving so much, wringing every drop from mysterious but wonderful existence, seeing the growing-up not just of one's children and grandchildren but also *their* children and grandchildren. Oh, yes please!

But wait. Our life span is our life span for a reason. Life spans vary enormously in the biological world, from barely a day in the mayfly to more than 100 years in the Galapagos tortoise and an estimated 1,500 years in the Antarctic sponge. These spans have been imprinted by natural selection because they are those that serve the species best—that maximize the tradeoff between caring for and competing with one's offspring.

Most of us love our parents. But imagine a world inhabited not only by your own parents but also everyone else's, and also your and their grandparents and your and their great-grandparents—a society run by people whose ideas and attitudes date back four centuries. Imagine a world in which your boss might be, for the next 100 years, in the post you covet. The generations would be competing with each other for food, housing, jobs, space. As it is, the young complain about how their elders, with their already rapidly increasing life spans, are driving up house prices by refusing to downsize in middle age, and driving up unemployment by refusing to retire. Imagine four centuries of people ahead of you in the housing and job queues.

The prolonging of the human life span is often lauded in the media, but it is almost never questioned. Nobody seems to doubt that we should push forward with aging research, identify those genes, tinker with them, make them work for us. For nobody wants to die, and so we all want this research to succeed. We want it for ourselves and our families. We want ourselves and our loved ones to live as long as possible—forever, if we can.

But is it the best thing for our species? Have almost 4 billion years of evolution been wrong? We are not Antarctic sponges or blue-green algae—we die for a reason. We die so that our youth—those better versions of ourselves—can flourish. We should worry about the loss of death.

GLOBAL GRAYING

DAVID BERREBY
Journalist & contributor, The New Yorker, The New York Times Magazine, Nature*; author,* Mind Matters *blog for bigthink.com; author,* Us and Them: The Science of Identity

What worries me is the ongoing graying of the world population, which is uneven globally but widespread. It is not on the radar (except for occasional gee-whiz news stories and narrow discussions about particular problems for this or that trade). But it should be—both the coming vast increase in the number of elderly people and the general rise in average age, as middle-aged and older people come to represent a greater share of humanity.

For example, out of the 9 billion people expected when the Earth's population peaks in 2050, the World Health Organization expects 2 billion—more than one person in five—to suffer from dementia. Is any society ready for this? Is any really talking about how to be ready?

At the coming mid-century, in rich nations, nearly one person in three will be more than sixty years old. But this upheaval won't be confined to the developed world. The Chinese population's median age is now almost thirty-five; by 2050 it will be forty-nine. India's population of people aged sixty to eighty will be 326 percent larger in 2050 than it is now. Elderly people, now 7 percent of Brazil's population, will make up nearly a quarter of that country in 2050. Yes, a swath of poor nations in Africa and Asia will soon go through classic population explosions and will teem with young people. But they will be exceptions to a global trend. "Before 2000, young people always outnumbered old people,"

Rockefeller University's Joel E. Cohen wrote a few years ago. "From 2000 forward, old people will outnumber young people."[b]

Awareness of this demographic shift is partial and piecemeal. Public health specialists discuss the expected huge increases in cases of "gray" diseases—chronic noncommunicable ailments like heart and lung problems, stroke, diabetes, and kidney failure. Economists talk about the disruptions that follow when working-age people are too few to support the retirees. Financial types bemoan the many millions of people who could not or would not lay away money and now face decades with no obvious way to get the income they'll need. Governments in India and China have introduced laws to prop up family values in the wake of stories of old people abandoned by their adult children. Within each discipline and profession there's some discussion of this vast disruption, but almost no one, to my knowledge, is discussing the underlying cause or trying to map out the consequences and relate them to each other.

The greatest worry about this shift turns on the social safety net. Most developed and developing nations promise at least some security and medical care for the old. That promise depends on the pyramid structure of a 20th-century society, in which active working-age people outnumber the retirees. It's hard to see how those social security guarantees can stay in place when there are fewer young workers for more and more older dependents. China, for example, now has about six workers for each retired person. On current trends, that ratio will be two workers per retiree by 2050.

That's a prescription for labor shortages, falling production, and political uproar as promised pensions and health care for the elderly become impossible to pay for. And this is one reason why China is expected to relax the famous "one-child" policy that now applies to a large proportion of its people—and why more and more nations are trying to increase their fertility rates. How else do you forestall this kind of crisis? Well, you could raise the retirement age so that workers support retirees for longer. But that's not acceptable for a number of practical and political reasons, one of which is that in a number of countries the retirement age would have go into the late seventies for this to work.

Politically, I worry about the consequences of a shift in power and influence away from younger people to the middle-aged and the old. In democracy, there is power in numbers, and if the numbers empower older people, then I fear that their concerns (for keeping what they have, for preserving the past as they imagine it, for avoiding the untried and unfamiliar) will start to overwhelm those of younger people. This has never happened before, so we can't know exactly what its consequences are. But I doubt they're good.

I think we can expect to see some frank and ugly intergenerational conflict. Even the United States, which faces less graying than other rich nations due to its openness to immigration, is in the beginnings of a policy debate about pensions and medical care that pits the interests of older workers (secure and predictable pensions and medical care) against those of younger people (education, future infrastructure, opportunity).

In countries where graying is happening really fast (Spain, Italy, Japan), one consequence might be an upsurge in xenophobic nationalism, for two reasons. First, there's good evidence that openness to change and new experiences declines with age. Robert Sapolsky devised some tests for this back in the 1990s and concluded that the window for being willing to try new music closed, for his American subjects anyway, at thirty-five and openness to new kinds of food ended around thirty-nine. People who might respond to a crude nationalist message—"Let's get back to the way things were in the old days, before things went bad!"—will be a bigger proportion of voters.

Second, because of the economic trouble I've already mentioned, nations will look for ways to boost their active workforces to support all those retirees. And here the options are (a) boost the birthrate, or (b) open doors to immigrants from all those poor, youthful nations in Africa and Asia, or (c) make a lot of robots (as they seem to be trying to do in Japan). Options (a) and (b) are obvious triggers of xenophobic rhetoric ("Young women must do their duty and make more of us!" and "We have too many of these damn foreigners now!").

Finally, on the cultural front, I worry about too much deference being given to the fears of people my age (fifty-four) and older. In the past twenty years, it has become intellectually respectable to talk about immortality as a realistic medical goal. I think that is an early symptom of a graying population. Here's another: When we speak of medical care, it's often taken as a given that life must be preserved and prolonged. Ray Kurzweil has said that whenever he asks a hundred-year-old if she wants to reach a hundred and one, the answer is yes. Has he asked the hundred-year-old's children, neighbors, or employees? They might give a different answer, but in a graying world the notion of natural limits and a fair share of life becomes taboo.

Economist and jurist Richard Posner described the pre-graying dispensation this way: "In the olden days, people broke their hips and died, which was great; now they fix them." Posner grew up with the simple and once commonsensical notion that you have your time on Earth and then you get out of the way. Because other, younger people are here, and they need to take your place. They'll need money for something other than their parents' practical nurses, they'll need the jobs older people won't retire from, they'll need the bandwidth that we old people will fill with talk of tummy tucks and Viagra. And they will need to have some decades out of the shadow of their parents. Once, people who were dutiful if not lovingly devoted to their elders could count on liberation. (I think of Virginia Woolf, whose father died at seventy-one, writing twenty-five years later: "He would have been 96, 96, yes, today . . . but mercifully was not. His life would have entirely ended mine. What would have happened? No writing, no books—inconceivable.") I fear it is becoming acceptable, due to the demographic shift, to tell younger generations that their day might never, should never, come.

Perhaps, though, I should be more positive and less like a typical fifty-four-year-old, extolling the way things were in my youth. No doubt the grayer world will have its advantages. Older people, for example, consume less power and fewer products per capita. In 2010, climate scientist Brian O'Neill and his colleagues analyzed that effect and concluded that global graying might supply as much as 29 percent of the reductions in carbon-dioxide emissions needed to avert a climate catastrophe this century. Whatever one's attitude toward graying, though, it is indisputably happening. It deserves much more attention.

ALL THE T IN CHINA

ROBERT KURZBAN
Psychologist; director, Pennsylvania Laboratory for Experimental Evolutionary Psychology (PLEEP), University of Pennsylvania; author, Why Everyone (Else) is a Hypocrite: Evolution and the Modular Mind

In 2020, by some estimates, there will be 30 million more men than women on the mating market in China, leaving perhaps up to 15 percent of young men without mates.

Anthropologists have documented a consistent historical pattern: When the sex ratio skews in the direction of a smaller proportion of females, men become increasingly competitive and more likely to engage in risky, short-term-oriented behavior, including gambling, drug abuse, and crime. This sort of pattern fits well with the rest of the biological world. Decades of work in behavioral ecology has shown that in species in which there is substantial variation in mating success among males, males compete especially fiercely.

The precise details of the route from a biased sex ratio to antisocial behavior in humans is not thoroughly understood, but one possible physiological link is that remaining unmarried increases levels of testosterone—often simply referred to as "T"—which in turn influences decision making and behavior.

Should all this T in China be a cause for worry?

The differences between societies that allow polygyny and those that don't are potentially illustrative. In societies with polygamy, there are, for obvious reasons, larger numbers of unmarried men than in societies that prohibit polygyny. These unmarried men compete for the remaining unmarried women;

they evince a greater propensity to violence and engage in more criminal behavior than their married counterparts. Indeed, cross-national research shows a consistent relationship between imbalanced sex ratios and rates of violent crime. The higher the fraction of unmarried men in a population, the greater the frequency of theft, fraud, rape, and murder. The size of these effects are nontrivial: Some estimates suggest that marriage reduces the likelihood of criminal behavior by as much as one half.

Further, relatively poor unmarried men historically have formed associations with other unmarried men, using force to secure resources they otherwise would be unable to obtain.

While increasing crime and violence in Asian countries with imbalanced sex ratios is a reason to worry in itself, the issue is not solely the potential victims of crimes that might occur because of the sex-ratio imbalance. Evidence indicates that surpluses of unmarried young men have measurable economic effects, lowering per-capita GDP.

China, of course, plays a crucial role in the modern heavily interconnected world economy and is the largest or second largest trading partner for seventy-eight countries. Although many Americans worry about China "overtaking" the United States—as if economics were a zero-sum game—the real danger stems from the ripples of a potential Chinese economic slowdown, whether from civil unrest or otherwise. Regional economies, such as South Korea and Taiwan, would no doubt be hard hit, but Europe and the United States would suffer disruptions of both supply and demand, with unpredictable but possibly substantial economic consequences.

The route from unmarried men to global economic meltdown is perhaps a bit indirect, but the importance of China in the world economy makes such threats to stability something to worry about.

TECHNOLOGY MAY ENDANGER DEMOCRACY

HAIM HARARI

Physicist; former president, Weizmann Institute of Science; author, A View from the Eye of the Storm: Terror and Reason in the Middle East

Science is the source of numerous cures for medical, social, and economic issues. It is also an incredibly exciting and beautiful intellectual adventure. It leads to new technologies, which change our lives, often for the better. Could these technologies endanger the foundations of liberal democracy? This may sound crazy. Yet we should all worry about it. It is a real threat, which should concern every thinking person, if he or she believes that science can advance humanity and that democracy is the least bad system of governance.

A serious mismatch is gradually developing, step by step, between two seemingly unrelated issues: the penetration of science and technology into all aspects of our life, and liberal democracy as practiced throughout the free world. Intrinsically, science and technology are neither good nor bad; it is how we use them that may lead to far-reaching benefits or negative results. Their applications, often planned and deliberate, are sometimes unintended and accidental. The developing conflict between the consequences of modern technology and the survival of democracy is unintended but pregnant with great dangers.

Let us count seven components of this brewing trouble:

First, a mismatch of time scales. Many issues tackled by decision makers are becoming more complex—multidisciplinary, global, multigenerational. Education systems, research policy, social security, geopolitical trends, health insurance, environmental issues, retirement patterns—all have time scales of decades. The time lag from discussion to decision, execution, and consequences is becoming longer, thanks to our growing ability to analyze long-term global effects and to more years of education, work, and retirement for the average person. On the other hand, the time horizon of politicians is always the next election—anything between two and seven years. Modern technology, while producing longer time scales for the problems, creates instant online popularity ratings for reigning office holders, pressing them for short-term solutions. We live longer, but think shorter.

The second component is another type of time mismatch. Twitter, texting (or SMS, in European jargon), Internet comments or "talkbacks," and similar one-liners make the old superficial 60-second TV news item look like an eternity. But real public issues cannot be summarized by micro sound bites. This encourages extremism and superficiality and almost forces politicians to express themselves in the standard 140 characters of Twitter, rather than in 140 lines, or 140 pages of a decent position paper. The voting public is exposed only to ultra-brief slogans, while a younger generation is becoming *homo neo-brevis*: the next evolutionary phase of the human race, with a brief attention span, affinity for one-liners, and narrow fingers for the smartphone.

The third issue is the growing importance of science literacy and quantitative thinking for decision makers. Today's world introduces us to energy issues, new media, genetic manipulations, pandemic flu, water problems, weapons of mass destruction, financial derivatives, global warming, new medical diagnostics, cyberwars, intellectual property, stem cells, and numerous other transformations that cannot be handled by people who cannot comprehend scientific arguments accompanied by simple quantitative considerations. Unfortunately, the vast majority of senior decision makers in most democracies lack these rudimentary abilities, leading to gross errors of judgment and historic mistakes that will affect many generations to come. We need scientifically trained political decision makers.

The fourth is the fact that electability to high office requires talents unrelated to those required for governing and leading. Major countries often elect senior office holders with credentials that would normally not get them a job as the CEO of a minor company. The democratic process starts not with a proper job description but with an ability to charm TV viewers and appear either as "one of the guys" or as a remote, admired prince (or, even better, both). TV and other electronic media make sure that most voters never see the real person but only an image on the screen, augmented by all possible add-ons. A talent for speech delivery, including the ability to read from a teleprompter while appearing to improvise, is more crucial than experience, familiarity with global issues, and leadership.

The fifth danger is the mad rush for "transparency," enhanced by immediate Web dissemination of all revealed items. It is almost impossible to have a proper frank high-level discussion, weighing outside-the-box options before rejecting them, toying with creative ideas, and expressing controversial views, when every word spoken may appear within days on the screens of a billion computers and smartphones, summarized by one sentence and often out of context. It is impossible to write an honest recommendation letter or a thorough well-balanced evaluation of an organization or a project when confidentiality is compromised and public disclosure is idolized. Small wonder that talented and experienced people with proven abilities in other fields shy away from entering politics, when "transparency" threatens to destroy them. One fears that future elected and appointed senior officials will have to post the results and pictures of their latest colonoscopy on the Web in the name of transparency.

The sixth component, also amplified by technology, is the public desire for freedom—of speech, the press, information, academic discourse, and all the other freedoms guaranteed by a proper democracy. These, as well as other human rights, are indeed the pillars of democracy. But when carried to unacceptable extremes they may lead to grave distortions: Incitement for murder or genocide is allowed; pedophilia is acceptable; disclosure of life-endangering national security information is fashionable; equal time for creationism and evolution is demanded; protecting terrorists' and murderers' rights is advocated more vigorously than defending victims' rights. Numerous other outlandish situations never meant to be covered by basic human rights emerge. Although technology itself has not created these situations, the brevity of messages and their fast and wide dissemination, together with the ability to transmit all of the above across borders from backward dictatorships into democracies, turn our sacred human rights and civil freedoms into double-edged swords.

Finally, the seventh pillar of the sad mismatch between modern technology and democracy is globalization. Political boundaries may define a state, a country, or a continent. But every political unit must have a certain set of rules. Country A can be an exemplary democracy and country B a dark dictatorship. If there is very little cross-talk between their societies, both regimes may survive and live by their own rules. Globalization helps to spread progressive ideas into dark political corners, but if in Germany the denial of the holocaust is a criminal offense, and a satellite transmission from Iran can reach directly every house in Germany, we have a new situation. If modern technology allows fast and efficient money laundering, performed by numerous international banks almost at the speed of light, we have a new challenge. If the world tries to make international decisions and treaties by majority votes of countries most of which have never experienced anything remotely similar to democracy, it enforces global antidemocratic standards. We also observe an enhancement of illegal immigration patterns, cross-boundary racist incitement, international tax evasion, drug trafficking, child labor in one region producing goods for another region that forbids it, and numerous phenomena amplified by the fast mobility and modern communication offered by today's technology.

To be sure, all of these seven points have been with us for years. We have often had short-sighted leaders, complained about superficial TV coverage of complex issues, observed scientifically illiterate leaders moving blindly in a labyrinth of technical issues, elected inexperienced good-looking politicians, demanded a reasonable level of transparency, exaggerated the application of honored constitutional principles, and believed in connecting with other nations on our planet. *But modern technology has changed the patterns of all of these and amplified a dangerous lack of balance between our ideals and today's reality.*

As someone who believes in the enormous positive contribution of science and technology for our health, food, education, protection, and understanding of the universe, I am in great pain when I observe these features, and I believe we should be truly worried. The only way to cope with the problem is to allow the structure of modern liberal democracy to evolve and adapt to the new technologies. That has not begun to happen. We do not yet have solutions and remedies, but there must be ways to preserve the basic features of democracy while fine-tuning its detailed rules and patterns so as to minimize the ill effects and allow modern science and technology to do more good than harm.

THE FOURTH CULTURE

BRUCE PARKER
Visiting professor, Center for Maritime Systems, Stevens Institute of Technology; author, The Power of the Sea: Tsunamis, Storm Surges, Rogue Waves, and Our Quest to Predict Disasters

If *Edge* is the Third Culture (scientists and other thinkers communicating new ideas about the world directly to the general public), and if the rest of the scientists and the so-called literary intellectuals are the first two cultures, as proposed by C.P. Snow, then there is also another culture whose impact is growing fast enough that it can justifiably be called the Fourth Culture. This "culture" is not really new.

In the past, we would simply have called it Popular Culture and then dismissed it as a world of mostly superficial entertainment with only a certain segment of the population caught up in it (that segment not considered intellectual or influential). But now this Popular Culture is Internet-driven and global and as a result has become pervasive, and its growing influence does not allow us to dismiss it so easily.

The Internet (and its associated media/communication entities, especially cell phones and cable TV) empower the Fourth Culture. And that culture is no longer concerned only with pop music, movies, TV, and video games. It now includes religion and politics and almost everything that touches people in everyday life. It is a bottom-up "culture" with a dumbing-down effect that is likely to have repercussions.

We should be worrying about a growing dominance by the

Fourth Culture and how it may directly or indirectly affect us all. Because of its communication capabilities and its appeal to people's egos, their sexuality, prejudices, faith, dreams, and fears, the Fourth Culture can easily shape the thoughts of millions. It promotes emotion over logic, self-centeredness over open-mindedness, and entertainment value and money-making ability over truth and understanding. And for the most part it ignores science.

The primary use of the Internet is still for entertainment, but that alone is a matter of concern. As more and more of the population fills more and more hours of the day with entertainment, this leaves fewer hours for activities that promote intelligence, compassion, or interest in anything that falls outside their own Internet-dominated microcosms. When one's "accomplishments" in life and self-image become focused on things like scoring the most kills in a video war game or being able to see one's favorite rock star in person or having one's favorite sports team win a game (all possible before the Internet but now carried to much greater extremes), what passion is left for the real world, for a job, or for the problems of fellow human beings? Would it be taking the point too far to suggest a parallel with the Romans, who kept the masses distracted from real-world problems by enticing them into the Colosseum to watch such spectacles as gladiators battling to the death?

The Fourth Culture is probably not a threat to science; too much money is being made from science (and its resulting technology) for science to disappear. We are not trying to improve the science and math scores of our students so we can produce more scientists. There will always be those children born with the boundless curiosity about how the world works that leads them into science or some other analytical type of work. The reason we want to do a better job of educating our children in science is to make them better citizens. Our citizens need to see the world and its problems through the eyes of science. They must be able to recognize logical approaches to problem solving and not be blinded by religious views, myths, or bogus fears promoted by opponents to a particular course of action.

Nowhere does the Fourth Culture cause more concern than in how it affects whom we select as our leaders, using a now-handicapped democratic system. How the products/results of science are used is now often in the hands of decision makers who do not understand science. Some of them think science is bad and that carefully studied and proved theories are no different than religious dogma.

We want elected officials to be intelligent people who care about doing what's best for all the people of the nation (or the world). But elected government positions are the only jobs that have no required criteria that prospective candidates must meet, other than a minimum age and, for a president, being born in the United States. Candidates for elected office do not need to have a college degree, or success in business, or any verifiable achievements in order to be elected. They simply need to convince people to vote for them, by whatever emotional means their campaign teams can come up with. Thanks to the Fourth Culture, it has become easier to elect uninformed and even stupid candidates, through emotional manipulation in the form of appeals to religion, patriotism, class distinctions, ethnic biases, and so on (appeals fueled by huge amounts of money, of course). Sound bites and campaign ads that look like movie trailers win out over carefully thought-out logical discourse.

The press, once called the fourth branch of a democratic government because it kept the other three branches honest, is now just "the media" and has distressingly lost much, or even most, of its watchdog capabilities. In an attempt to survive financially in this Internet–dominated media world, the press has cut back newsrooms, relied more on unsubstantiated sources from the Internet, treated pop stories as news (reducing the space devoted to important stories, especially scientific stories), and allowed even the most idiotic and abusive comments to be left on their Web sites (in the name of free speech, but really to have as many comments as possible to prove to advertisers that their Web sites are popular).

We had hoped that the Internet would be a democratizing force. We had hoped that it would give everyone a voice and bring to light new ideas and new approaches to solving serious problems. To some degree it has, and it still may do a lot more. But the Internet has also given a voice to the ignorant. A voice they never had before. A loud and emotional voice. We can hope that the effects of the growing influence of the Fourth Culture do not turn out to be destructive on a large scale, but it is something worth worrying about, and it is worth looking for ways to reduce its impact. This is one more reason why expanding the Third Culture is so important.

CLASSIC SOCIAL SCIENCES' FAILURE TO UNDERSTAND "MODERN" STATES SHAPED BY CRIME

EDUARDO SALCEDO-ALBARÁN
Philosopher & political scientist; director, Scientific Vortex, Inc.

According to classic political science and history, after medieval kingdoms based on the will of God and the king, modern secular and liberal states arose. Those modern states are sustained by secular laws intended to protect social welfare and individual autonomy. Apparently most states in the West have adopted democracy, impartial laws, and protection of human rights and therefore have moved toward modernity. However, we should worry about several "modern" states that, in practical terms, are shaped by crime—states in which laws are promulgated by criminals and which, even worse, have been legitimized through formal and "legal" democracy. The scare quotes are there because, usually, the social reality is not what the classic social sciences show it to be.

Crime is commonly thought to be, as in the movies, about bad guys confronting good guys; the two only sporadically get together, through bribery, say, or infiltration. However, data show that a gray area of collaboration and co-optation among good guys and bad guys is a constant: insurgents and counterinsurgents, public servants, politicians, candidates, and various private agents work together to define the rules of societies and shape institutions according to their own (sometimes criminal) interests.

In such states, most of the trappings of formal democracy are preserved: Bills are proposed, laws are approved, and electoral processes are carried out. Beyond this formalism, members of criminal networks are the actors in those processes. In African countries such as Sudan or the Democratic Republic of the Congo, private industry, insurgents, and public servants engage in a circular causality of corruption conducive to enormous violence.

Similar circumstances obtain in Mexico, Guatemala, and Colombia. In 2012, the United States requested the extradition of a former Guatemalan president for participating in a large-scale money-laundering scheme. In Mexico, in 2009, thirty-eight mayors of the State of Michoacán were arrested for collaborating with the criminal network La Familia. In Colombia, in July 2001, thirty governors, mayors, political leaders, and candidates signed a secret pact with the counterinsurgency group United Self-Defense Forces of Colombia to reestablish the Colombian state by creating a new set of rules. Some of the signers then became national legislators.

This is not a matter of sporadic corruption among enforcement officers; this is a matter of massive corruption and co-optation in the defining of laws and institutions—the same laws and institutions that sustain states. This is about situations in which the "bad guys" no longer confront the "good guys"; instead, they become one with the formal state, and their criminal interests are protected by the state's laws and institutions.

Several questions arise: What can be said about laws proposed and approved by criminals? Are those "real" laws? Shall we ask individuals to decide to obey or disobey a particular law depending on whether or not the legislators favor criminal groups? What about those cases in which the legislators favor not criminal groups but powerful corporations? Do we need a different definition of "law"? Do we need a different definition of "state"?

This is not only a conceptual issue.

Civil and penal laws are at the foundation of everyday life.

What we think and how we behave is the result of interaction between our brains and the codes, rules, and institutions around us. When those codes and institutions are based on criminal interests, the social and cultural definitions of right and wrong change. Then we have a society in which illegality is the norm—a society usually defined as a failed state. The result is a vicious circle that strengthens criminal networks operating across different countries—networks involved in widespread corruption, mass murder, human trafficking, kidnapping, extortion, and violence. The classic social sciences—with their dichotomies, such as legal/illegal, rational/emotional, micro/macro, and individual/social, among others—are not equipped to understand and explain the current social reality in such countries.

Beyond the traditional qualitative and quantitative tools, new integrative approaches are needed to explain current social reality. We should therefore also worry about the failure of social scientists to call attention to situations like those I have just described. While they focus on explaining that reality by means of their classic dichotomies and frameworks, some states are deteriorating into chaos.

Just as the convergence and elimination of boundaries in medicine, engineering, genetics, artificial intelligence, and biochemistry have redefined the concept of life through new and integrative approaches, the convergence and elimination of boundaries between the various social sciences will allow the redefinition of concepts like "crime," "law," and "state," making them more consonant with reality. Just as we can no longer see the world in terms of organic/inorganic or natural/artificial, we can no longer see the social world in terms of legal/illegal or emotional/rational.

We need to worry both about “modern” states shaped by crime and about the classic scientists trying to explain them. Society needs armies, enforcement, and public policy, but even more important, society needs an army of challenging and *Edg*y forward thinkers who are not afraid of eliminating boundaries between sciences and care more about improving social reality than publishing in respected journals—an “army” of *Edgies* who help weak states achieve the great potential of humankind.

IS THE NEW PUBLIC SPHERE . . . PUBLIC?

ANDREW LIH

Associate professor, Annenberg School for Communication & Journalism, University of Southern California; author, The Wikipedia Revolution

The advent of social-media sites has allowed a new digital public sphere to evolve by facilitating many-to-many conversations on a variety of multimedia platforms, from YouTube to Twitter to Weibo. It has connected a global audience and provided a new digital commons that has had profound effects on civil society, social norms, and even regime change in the Middle East. As important as it has become, are critical aspects of this new public sphere truly public?

There are reasons to be worried.

While we are generating content and connections that are feeding a rich global conversation unimaginable just ten years ago, we may have no way to re-create, reference, research, and study this information stream after the fact. The spectrum of challenges is daunting, whether it's because information is sequestered in private hands, kept from full access, deleted from sight, retired with failed businesses, or shielded from copying because of legal barriers.

Twitter, in particular, has emerged as the heart of a new global public conversation. However, anyone who has ever used its search function knows that the second chance to find content is dubious. Facebook works in a private eyes-only mode by default and is shielded even more from proper search and inspection, not only by the public but even by the creators of the original content.

How about the easier case of individuals simply asserting control over their own content within these services? Users of social-media content systems still have sole copyright of their content, though the terms of service that users agree to is rather extensive. Twitter's is fairly typical: "You grant us a worldwide, non-exclusive, royalty-free license (with the right to sublicense) to use, copy, reproduce, process, adapt, modify, publish, transmit, display and distribute such Content in any and all media or distribution methods (now known or later developed)."

Without passing judgment on the extent or reach of these types of license agreements, the logistics of accessing one's own data are worrisome. Typically, these services (Twitter, Facebook, Weibo, Instagram, *et al.*) are the sole digital possessors of your words, links, images, or videos created within their systems. You may own the copyright, but do you actually possess a copy of what you've put into their systems? Do you actually control access to your content? Do you have the ability to search and recall the information you created? Is public access to your data (e.g., through application programming interfaces) possible now, or guaranteed in the long term?

That we continue to use an array of information systems without assurances about their long-term survivability or commitment to open access, and without knowing whether or not they are good stewards of our history and of public conversation, should worry us all.

What can be done about this?

To its credit, Twitter has partnered with the Library of

Congress to hand over the first four years' worth of tweets, from 2006 to 2010, for research and study. Since that first collaboration, it has agreed to feed all tweets to the library on an ongoing basis. This is commendable, but it's drinking from a virtual firehose, with roughly half a billion new tweets generated every day. Few entities have the technology to handle very big data, and this is truly massive data.

The Twitter arrangement has provided quite a challenge to the library, as they don't have an adequate way to serve up the data. By their own admission, they haven't been able to facilitate the 400 or so research inquiries for this data, because they are still addressing "significant technology challenges to making the archive accessible to researchers in a comprehensive, useful way." So far, the library hasn't planned on allowing the entire database to be downloaded in its entirety for others to have a shot at crunching the data.

We have reasons to worry that this new digital public sphere, while interconnected and collaborative, is not a true federation of data that can be reconstructed for future generations and made available for proper study. Legal, infrastructural, and cooperative challenges abound that will likely keep it fractured, perforated, and incoherent for the foreseeable future.

BLOWN OPPORTUNITIES

FRANK WILCZEK

Herman Feshbach Professor of Physics, MIT; 2004 Nobel laureate in physics; author, The Lightness of Being: Mass, Ether, and the Unification of Forces

We should be afraid, above all, of squandering our grand opportunities.

Today many hundreds of millions of people enjoy a material standard of living higher than any but a fortunate few attained just a century ago. More than that: Things we take for granted today, such as instant access to distant friends and to the world's best artistic and intellectual productions, were barely imaginable then.

How did that miracle happen?

The enabler was new basic knowledge. Insight into electromagnetism allowed us to transmit both power and information; first steps in understanding the quantum world have supported microelectronics, lasers, and a host of other technologies. Understanding the true origin of diseases allowed their prevention or cure; molecular understanding of plant and animal metabolism supported vast improvements in agriculture, yielding richer harvests with far less human labor.

Fundamental breakthroughs, by their nature, cannot be predicted. Yet there is every reason to hope for more, soon. The tools of quantum science and molecular biology are sharper than ever, and now they can be harnessed to extraordinary, ever-strengthening computer power. Opportunities beckon: the challenge of making quantum computers, real possibilities for enhancing human minds and increasing the span of healthy life, and others not yet imagined.

What could go wrong?

Gibbon, in describing the fall of Rome, spoke of "the natural and inevitable effect of immoderate greatness" and of "the triumph of barbarism and religion." Those forces—the diversion of intellectual effort from innovation to exploitation, the distraction of incessant warfare, rising fundamentalism—triggered a Dark Age before, and they could do so again.

THE POWER OF BAD INCENTIVES

SAM HARRIS
Neuroscientist; chairman, Project Reason; author, Free Will

Imagine that a young white man has been falsely convicted of a serious crime and sentenced to five years in a maximum security penitentiary. He has no history of violence and is understandably terrified at the prospect of living among murderers and rapists. As the prison gates shut behind him, a lifetime of diverse interests and aspirations collapses to a single point: *He must avoid making enemies so that he can serve out his sentence in peace.*

Unfortunately, prisons are places of perverse incentives—in which the very norms one must follow to avoid becoming a victim lead inescapably toward violence. In most U.S. prisons, for instance, whites, blacks, and Hispanics exist in a state of perpetual war. This young man is not a racist and would prefer to interact peacefully with everyone he meets, but if he does not join a gang he is likely to be targeted for rape and other abuse by prisoners of all races. To not choose a side is to become the most attractive victim of all. Being white, he likely will have no rational option but to join a white supremacist gang for protection.

So he joins a gang. In order to remain a member in good standing, however, he must be willing to defend other gang members, no matter how sociopathic their behavior. He also discovers that he must be willing to use violence at the tiniest provocation—returning a verbal insult with a stabbing, for instance—or risk acquiring a reputation as someone who can be assaulted at will. To fail to respond to the first sign of disrespect with overwhelming force is to run an intolerable risk of further abuse. Thus the young man begins behaving in precisely those ways that make every maximum security prison a hell on Earth. He also adds further time to his sentence by committing serious crimes behind bars.

A prison is perhaps the easiest place to see the power of bad incentives. And yet in many other places in our society, we find otherwise normal men and women caught in the same trap and busily making life for everyone much less good than it could be. Elected officials ignore long-term problems because they must pander to the short-term interests of voters. People working for insurance companies rely on technicalities to deny desperately ill patients the care they need. CEOs and investment bankers run extraordinary risks—both for their businesses and for the economy as a whole—because they reap the rewards of success without suffering the penalties of failure. Lawyers continue to prosecute people they know to be innocent (and defend those they know to be guilty) because their careers depend upon winning cases. Our government fights a war on drugs that creates the very problem of black market profits and violence that it pretends to solve. . . .

We need systems that are wiser than we are. We need institutions and cultural norms that make us better than we tend to be. It seems to me that the greatest challenge we now face is to build them.

SCIENCE PUBLISHING

MARCO IACOBONI

Neuroscientist; professor of psychiatry & biobehavioral sciences, David Geffen School of Medicine, UCLA; author, Mirroring People: The Science of Empathy and How We Connect with Others

We should be worried about science publishing. When I say "science publishing," I am really thinking about the peer-reviewed life-science and biomedical literature. We should be worried about it because it seems that the only publishable data in life science and biomedical literature are novel findings. That's a serious problem, because one of the crucial aspects of science is reproducibility of results. The problem in life science is that if you replicate an experiment and its results, no one wants to publish your replication data. "We know that already," is the typical response. Even when your experiment is not really a replication but resembles a previously published one, and your results aren't even exactly identical to previously published ones but close enough, unless you find a way of discussing your data under a new light, nobody wants to see your study published. Only experiments that produced results opposite those of previously published studies are likely to be published. Here, the lack of replication makes the experiment interesting.

The other big problem is that experiments that produce negative findings, or "null results"—that is, do not demonstrate any experimental effect—are also difficult to publish unless they show lack of replication of a previously published important finding.

These two practices combined make it very difficult to figure out, on the basis of the literature alone, which results are solid and replicable and which are not. And that's clearly a problem.

Some have argued that to fix this problem we should publish all our negative results and publish positive results only after replicating them ourselves. I think that's a great idea, although I don't see the life-science and biomedical community embracing it anytime soon. But let me give you some practical examples as to why things are messed up in the life-science and biomedical literature and how they could be fixed.

One of the most exciting recent developments in human neuroscience is what's called *noninvasive neuromodulation*. It consists of a number of techniques using either magnetic fields or low currents to stimulate the human brain painlessly and with no, or negligible, side effects. One of these techniques has already been approved by the Food and Drug Administration to treat depression. Other potential uses include reducing seizures in epileptic patients, improving recovery of function after brain damage, and in principle even improving cognitive capacities in healthy subjects.

In my lab, we are doing a number of experiments using neuromodulation, including two studies in which we stimulate two specific brain sites of the frontal lobe to improve empathy and reduce social prejudice. Every experiment has a rationale that is obviously based on previous studies and theories inspired by those studies. Our experiment on empathy is based mostly on our previous work on mirror neurons and empathy. Having done a number of studies ourselves, we are pretty confident about the background on which we base the rationale for our experiment. The experiment on social prejudice, however, is inspired by a clever paper recently published by another group that also used neuromodulation of the frontal lobe. The cognitive task used in that study shares similarities with the cognitive mechanisms of social prejudice. However, here is the catch: We know about that published paper (because it was published), but we have no idea whether a number of groups attempted to do something similar and failed to get any effect—simply because a negative findings don't get published. We also can't possibly know how replicable the study is that inspires our experiment, because replication studies don't get published either. In other words, we have many more unknowns than we would like.

Publishing replications and negative findings would make it much easier to know what is empirically solid and what is not. If twenty labs perform the same experiment, and eighteen get no experimental effects, while the remaining two get contrasting effects, and all these studies are published, then you know, simply from reading the literature, that there isn't much to be pursued in that line of research. But if fourteen labs get the same effect, three get no effect, and three get the opposite effect, it is likely that the effect demonstrated by the fourteen labs is much more solid than the effects demonstrated by the six other labs.

Given the current publishing system, achieving these conclusions will be complicated. One way of doing it is to pool experiments that share a number of features. For instance, our group and others have investigated mirror neurons in autism and concluded that mirror neuron activity is reduced in autism. Some other groups failed to demonstrate it. The studies showing mirror neuron impairment in autism largely outnumber the studies failing to show it. In this instance, it is reasonable to draw solid conclusions from the scientific literature. In many others, however, as in the example of neuromodulation of the frontal lobe and social prejudice, there is much uncertainty, because of the selectivity regarding what gets published and what doesn't.

The simplest way to fix the problem is to evaluate whether or not a study should be published only on the basis of the soundness of its experimental design, data collection, and analysis. If the experiment is well done, it should be published whether it is a replication or not, no matter what kind of results it shows. A minority in the life-science and biomedical community are finally voicing this alternative to the current dominant practices in scientific publishing. If this minority eventually becomes a majority, we will at last have a scientific literature that can be evaluated in quantitative terms (*x* number of studies show this, while *y* number of studies show that) rather than in qualitative terms (this study shows *x* but that study shows *y*).

This approach will make it even more difficult for irrational claims (denial of evolution or denial of climate change are the most dramatic examples) to pretend to be "scientific." It would also limit the number of controversies in life science to those issues that are truly unclear, saving all of us the time we spend arguing about questions that should long ago have been settled by the empirical data.

EXCELLENCE

ERIC R. WEINSTEIN
Mathematician and economist; managing director, Thiel Capital

Over the past two decades I have been involved with the war on excellence.

I know that those few of us involved in the struggle are deeply worried about the epidemic of excellence, precisely because excellence compels its hosts to facilitate its spread by altering their perception of its costs and benefits. Most educated people have come to revere the spending of the fabled "10,000 hours" in training to become respected jacks of one trade. Large numbers of Americans push their inquisitive children away from creative play so that they can excel in their studies, in hopes that they will become excellent candidates for admission to a center of excellence and join the pursuit of excellence upon graduation.

The problem with all this is that we cannot excel our way out of modern problems. Within the same century, we have unlocked the twin nuclei of both cell and atom and created the conditions for synthetic biological and even digital life with computer programs that can spawn both descent and variation on which selection can now act. We are in genuinely novel territory, which we have little reason to think we can control; only the excellent would compare these recent achievements to harmless variations on the invention of the compass or steam engine. So, surviving our newfound godlike powers will require modes that lie well outside expertise, excellence, and mastery.

Going back to Sewall Wright's theory of adaptive landscapes of fitness, we see four modes of human achievement paired with what might be considered their more familiar accompanying archetypes:

A) ***Climbing—Expertise:*** Moving up the path of steepest ascent toward excellence for admission into a community that holds and defends a local maximum of fitness.
B) ***Crossing—Genius.*** Crossing the "Adaptive Valley" to an unknown and unoccupied even higher maximum level of fitness.
C) ***Moving—Heroism.*** Moving "mountains of fitness" for one's group.
D) ***Shaking—Rebellion.*** Leveling peaks and filling valleys for the purpose of making the landscape more even.

The essence of genius as a modality is that it seems to reverse the logic of excellence. Sometimes we must, at least initially, move away from apparent success and headlong into seeming failure to achieve outcomes few understand are even possible. This is the essence of the so-called Adaptive Valley, which separates local hills from true summits of higher fitness. Genius, at a technical level, is the modality combining the farsightedness needed to deduce the existence of a higher peak with the character and ability to survive the punishing journey to higher ground. Needless to say, the spectacle of an individual moving against his or her expert community, away from carrots and toward sticks, is generally viewed as a cause for alarm regardless of whether that individual is a malfunctioning fool or a genius about to invalidate community groupthink.

The heroes and rebels don't even accept the landscape as immovable but see dunes of fitness to be shifted by a sculpting or leveling of the landscape, with an eye toward altering the fitness of chosen populations.

None of these modes is intrinsically good or bad. Successful individuals generally maintain a portfolio of such modalities dominated by a leading chosen modality. But the first mode of excellence-driven expertise has, with institutional support, transformed into something unexpected, like a eusocial networked supercompetitor crowding out genius and heroism for institutional support within the research enterprise. An obviously stupid idea, like "self-regulating financial markets," now spreads frictionlessly among fungible experts inhabiting the now interoperable centers of excellence within newspapers, government, academe, think tanks, broadcasting, and professional associations. Before long, the highest levels of government are spouting nonsense about "the great moderation" in front of financial disaster.

I have searched almost the entire landscape of research and been shocked to find excellence having almost universal penetration, with the possible exceptions of Silicon Valley and hedge funds. Excellence, as the cult of history's second string, brooks no argument. As a pathogen, it spreads quickly, as if a virus, preferring to lie in wait on Ivy League sheepskin or other vectors of infection.

In the past, many scientists lived on, or even over, the edge of respectability, with reputations as skirt-chasing, hard-drinking, bigoted, misogynistic, childish, slutty, lazy, politically treacherous, incompetent, murderous, meddlesome, monstrous, and mentally unstable individuals, such as (respectively) John von Neumann, George Gamow, William Shockley, James Watson, Albert Einstein, Marie Curie, Stephen Smale, J. Robert Oppenheimer, Francis Crick, Paul Ehrenfest, Serge Lang, Edward Teller, and Alexander Grothendieck, who fueled such epithets with behaviors indicating that they cared little for what even other scientists thought of their choices.

But such disregard, bordering on deviance and delinquency,

was often outweighed by feats of genius and heroism. We have spent the last decades inhibiting such socially marginal individuals or chasing them out of our research enterprise and into startups and hedge funds. As a result our universities are increasingly populated by the over-vetted specialist and becoming the dreaded centers of excellence that infantilize and uniformize the promising minds of greatest agency.

If there is hope to be found in this sorry state of affairs it is in the rise of an archipelago of alternative institutions alongside the assembly line of expertise. This island chain of mostly temporary gatherings has begun to meet the need for heroism and genius. The major points of the archipelago are heterogeneous compared to their ivy-covered counterparts and include Burning Man, Foo Camp, TED, Breakout Labs, *Edge*, Sci Foo, Y Combinator, the Thiel Fellowship program, INET, FQXi, and Summit Series, to name a few—as well as some that are even more secretive.

In the wake of the *Challenger* disaster, Richard Feynman was mistakenly asked to become part of the Rogers Commission investigating the accident. In a moment of candor, Chairman Rogers turned to Neil Armstrong in a men's room and said, "Feynman is becoming a real pain." Such is ever the verdict pronounced by steady hands over great spirits. But the scariest part of this anecdote is not the story itself but the fact that we are, in the modern era, now so dependent on old Feynman stories, having no living heroes with which to replace him: the ultimate tragic triumph of runaway excellence.

UNMITIGATED ARROGANCE

JESSICA L. TRACY

Associate professor of psychology, University of British Columbia

I worry about the recent epidemic of lying and cheating that has infected public discourse in diverse domains. Think of science writer Jonah Lehrer's fabrication of quotes in his 2012 book, *Imagine: How Creativity Works*—which was subsequently pulled from shelves by its publisher. Or social psychologist Diederik Stapel's fabrication of empirical data reported in more than fifty published articles—most of them eventually retracted by the journals. Or Lance Armstrong's years of competitive cycling powered by illegal doping, resulting in the removal of his seven Tour de France victories and a lifetime banishment from the sport.

These problematic behaviors resulted either from a technological advance or a shift in the social climate. The current mass appeal of social psychology and social-science literature created a high payoff for smart and creative people like Lehrer and Stapel, who were able to attain a level of fame from writing social science—a level until recently inconceivable. Armstrong was lucky (or unlucky) enough to come of age in cycling at the time when blood-doping technology became largely undetectable. But changes like these are only the proximate causes of the epidemic. There is a broader, deeper psychological cause, and it is far from recent; it has been part of human nature throughout our evolutionary history. The psychological mechanism motivating and facilitating these corrupt behaviors is *hubristic pride*—the arrogance and egotism that drive people to brag, lie, cheat, and bully others to get ahead.

Hubristic pride is distinct from the confident authentic pride we feel in well-earned achievements. While authentic pride motivates hard work, persistence, and empathic concern for others, hubristic pride motivates hostility, aggression, intimidation, and prejudice. And this makes sense, because feeling hubristically proud does not mean feeling genuinely good about oneself. Instead, it involves inflated, inauthentic, and superficial feelings of grandiosity, which are used strategically and defensively to compensate for deep-seated, often unconscious insecurities. The hubristically proud are narcissistic but have low self-esteem and are prone to shame. Arrogance is how they cope with, and hide, their suppressed self-doubts. And because any kind of pride feels better than shame, those who feel hubristic pride seek to maintain it at any cost—finding new ways of promoting themselves and derogating others. Like a drug, hubristic pride makes getting ahead feel essential, as the only way to keep insecurities at bay. But the insecurities occasionally bubble up to the surface of awareness, reminding the hubristically pride-prone that they are not good enough, smart enough, or fast enough and leaving them with no option but to go beyond what they can achieve on their own. They use force, aggression, lying, and cheating to maintain the power and pride they have come to depend on. And hubristic pride convinces them they can get away with it.

The evolution of hubristic pride, which underlies the universal human motivation to climb the social hierarchy, is nothing new. What is new is that the bullies who feel it have a bully pulpit. Lehrer and Stapel were not the first writers or scientists to seek fame, but they were working in a new climate, where science and science writing are a means of attaining fame. As for Armstrong, by coming back from a near fatal cancer to win the world's most difficult race seven times, he became the first professional cyclist to achieve the name recognition of a movie star.

What is the solution to my worry? Ideally, institutions will develop better ways to catch liars and cheaters and enforce more severe penalties against them, so that the risk-to-payoff balance tilts in the other direction. But there may be another solution. We cannot stop people from feeling hubristic pride; it's part of our human nature, and given that power provides financial and reproductive benefits, it is evolutionarily adaptive. But we can become alert to its presence and pitfalls and catch it earlier. Arrogance is obvious, and research in my lab has found that people quickly and accurately identify the most dominant members of their social group—the ones likely to feel the most hubristic pride. What's difficult is avoiding falling prey to their manipulative influence and calling them out instead. Is that even possible? Perhaps, but only if we start questioning the success stories that seem too good to be true. This means sacrificing the collective pride we feel in the apparent accomplishments (and even the arrogance) of our cultural heroes. By enabling others' arrogance, we nurture the pride that can lead to large-scale deception and even crime and further increase the gap between true accomplishments and just rewards.

THE DECLINE OF THE SCIENTIFIC HERO

ROGER HIGHFIELD
Director, external affairs, Science Museum Group; coauthor (with Martin Nowak), SuperCooperators: Altruism, Evolution, and Why We Need Each Other to Succeed

In post-Olympic Great Britain, everyone is still basking in the glory of summer 2012's crop of sporting superheroes. Thanks to Bradley Wiggins, Jessica Ennis, Mo Farah, and all the other British medalists, there's overwhelming support for the view that the games were a great value, despite the huge cost of around £9 billion. But by comparison, how many heroes of last year's science does anyone remember? Answer: British theoretical physicist Peter Higgs, and for work he did half a century ago.

This worries me, because science needs heroes for the same reason the Olympics does. If we abandon our heroes, we make science insipid. And if it's boring, science loses support and funding. We need science to inspire and engage ordinary people more than ever before, because, through technology, it's the most powerful force acting on today's culture. What a difference it would make to how science is regarded by the public if we had a few more contemporary scientific heroes.

But it's becoming harder to find them. One reason is that traditional accounts of scientific discovery have been so flawed that the very idea of a hero has fallen into disrepute. These accounts tend to elevate a few brilliant men into geniuses but consign the rest—including almost all women—to oblivion. Historians of science have made much of how the record has been egregiously distorted by individuals seeking to lionize disciplines and glorify nations. They emphasize that discovery is a story of many participants, not of lone heroes or "eureka" moments. I don't disagree. Scientists are engaged in a great cooperative venture, one in which they build on the hard work of their predecessors, their peers, and their competitors. The problem is that when taken to extremes, this view can give the impression that science is carried out by a vast and faceless army exploring the infinite parameter space of possible experiments or theory, where breakthroughs are inevitable and individual endeavor counts for little.

A second force undermining heroes comes from the relentless rise of Big Science. Some 10,000 visiting scientists, half of the world's particle physicists, come to CERN in Geneva to join forces in their research. The members of the International Human Genome Sequencing Consortium came from a vast array of institutions. The follow-up Encode analysis of the genome relied on more than 440 scientists. Even mathematics, traditionally a lonesome profession, is becoming more collaborative, thanks to Cambridge mathematician Tim Gowers' Polymath Project.

Because of the rise of collaborative science and the efforts of historians, are heroes doomed to go extinct?

I hope not. As Hollywood already knows, we are hardwired to appreciate narratives based on individuals. This was vividly shown in the 1944 study by psychologists Fritz Heider and Mary-Ann Simmel in which people were shown an animation of a pair of triangles and a circle moving around a square.[c] Participants told stories about the circle and the little triangle being in love, the big bad gray triangle trying to steal away the circle, how they embraced and lived happily ever after, and so on. Recently, brain scans by Chris and Uta Frith of University College London and Francesca Happé of Kings College revealed activation in the temporo-parietal junction and medial prefrontal cortex during these scenarios.[d] What is remarkable is that the same brain network is engaged whether considering moving shapes or the mental states of others. There is a human predilection to make narratives out of whatever we see around us, to see agency in dark shadows and messages in the stars. In these patterns we can find heroes, too. The stories that create heroes are important because they cement reputations, crucial for the evolution of cooperation, through a mechanism called indirect reciprocity.

Yes, the reality of the story of a scientific development is always more complex than heroic stories suggest. But we have to be pragmatic when it comes to presenting the "truth" of a scientific advance to a general audience. This is not a choice between telling people the complete, nuanced, complex story behind a breakthrough and telling tales. When it comes to the public, it is usually a choice between saying nothing of interest or giving an engaging, heroic account. When it comes to selling the magic of science, we need to accept that the most powerful way is through heroic stories.

Moreover, when taken to extremes the collectivist view can be trite. Every time I drink a great cup of coffee, I could thank the farmers in Colombia who grew the beans, those in Brazil who provided the lush green fields of swaying sugar cane used to sweeten it, or the herdsman in Devon who milked the cows so my pick-me-up could be decorated with a little froth. I could also thank the nuclear power workers who provided the electricity to heat it, the person who had the bright idea of drinking a beverage based on roasted seeds in the first place, or who patented the first espresso machine. I could list all those hundreds of people who worked in supply lines straddling the planet to bring the energy, information, and ingredients together. I prefer to thank the barista, since our amazing ability to cooperate is a defining characteristic of human society, and it's a given that many others were involved in almost everything we do.

You are going to complain that hero worship can distort our picture of the way science is done. Yes it can. But the public is sophisticated enough to know that real life is always more complicated than the spectacle of an athlete standing on the podium. Nobody pretends otherwise. In the Paralympics, the heroes routinely thanked their family, sponsor, trainers, friend who paid for driving lessons, sports psychologists, sports scientists, and so on for their gold medals. We all know that, but that does not diminish the impact of winning a gold.

You are going to complain that all I'm really saying is that we need heroes because the public is slow on the uptake. Not at all. It's just that I believe in talking to people in a language they understand and in a way that ensures they will be receptive. We need convenient truths that can convey the kernel of science to a general audience without exhaustive emphasis on the collective aspects of discovery, let alone recourse to calculus, jargon, and dense descriptions. Scientists routinely use metaphors to communicate complex ideas, and by the same token you need heroic characters as metaphors to convey the broad sweep of scientific developments. The point is that heroes work as viral transmitters of science in the crowded realm of ideas. That's important, because we need as many people as possible to know what science is about if modern democracy is to function.

That is why, ultimately, we should be worried about the decline of heroes of science. The culture of skepticism, testing, and provisional consensus-forming in scientific research is the most significant achievement of our species, and it's time that everyone understood that (OK, it's going to be hard when it comes to the politicians). The real issue is not whether or not we should have heroes—of course we should—but how to ensure that they tell a reasonably truthful story about that amazing and supremely important endeavor we know as science.

AUTHORITARIAN SUBMISSION

MICHAEL VASSAR

Cofounder & chief science officer, MetaMed Research; former executive director, Singularity Institute

No one really knows what we should be worried about; the smartest people around seem to generally think the answer is machine intelligence. They may be right, but other types of inhuman intelligence have a track record worth worrying about. I'm worried that so few of us are doing anything to fight back against it.

In a pre-verbal tribe, most of the monkeys don't need to decide where they're going—because if they go off on their own, either the leader will kill them for rebelling or a jaguar will kill them once they're separated from their pack. Only pack leaders need to activate the messy symbolic intuitions that tell them what's out there and what needs to be done. As Abraham Maslow might say, only people who see themselves as chiefs, who have satisfied their need for esteem, are in a position to self-actualize.

Maslow suggests that at one level of description a human being consists of five programs. One escapes immediate danger, one seeks comfort and physical security, one finds a social context in which to be embedded, one builds esteem within that context, and one directs big-picture intentionality. When a given program reports satisfaction, the next program is turned on and starts competing in its more subtle game.

Some of those programs attend to things that can be understood fairly rigorously, like a cart, a plow, or a sword. Others attend to more complicated things, such as the long-term alliances and reproductive opportunities within a tribe. The former programs might involve situational awareness and detailed planning, while the latter might operate via subtle and tacit pattern-detection and automatic obedience to crude heuristics. If so, the latter programs might be fairly easily hacked. They might also offer a less fertile ground for the emergence of a Universal Turing Machine. Mechanical metaphors of solidity and shape might constitute a good substrate for digital, and thus potentially abstract, cognition, while social metaphors for vague properties like weirdness, gravitas, and sexiness might constitute a poor foundation for universal cognition. These programs seem to have been disfavored by history's great scientific innovators, who tend to make statements like "I do not know what I may appear to the world, but to myself I seem to have been only like a boy playing on the seashore, and diverting myself in now and then finding a smoother pebble. . . ." or "What do you care what other people think?"—which sound like endorsements of physical over social cognition.

So I worry about the consequences of assisting a population in satisfying its physiological needs but not its love, belonging, and esteem needs. Such a population would be at risk for underdevelopment of the programs for satisfying physiological and safety needs. That would be harmful if these functions are the basis for precise thought in general. One might argue that the solution to this problem is to hurry people through their love, belonging, and esteem needs, but that approach incurs further hazards.

Dacher Keltner and other psychologists have showed that higher socioeconomic class—essentially, higher satisfaction of esteem needs—leads to increased unethical behavior; other research suggests that external efforts to boost self-esteem tend to produce antisocial narcissism. Research of this type fits in well with Jonathan Haidt's theories on the authoritarian dimension of moral cognition. It may not be a coincidence that one of the world's most popular religions is literally named "submission." In general, authoritarian impulses have always prompted rulers to keep their subjects' bellies full and pressed firmly against the ground. Steven Pinker has compellingly established the success of modern society in bringing us unprecedented peace. People who don't fear for their safety but who despair of ever achieving love or belonging are the most submissive.

When we say someone is "smart," we largely mean that they learn quickly. In circumstances where most students reach a seventh-grade reading level, the "gifted" students absorb the full content of their schooling. According to New York City's three-time Teacher of the Year John Taylor Gatto, that curriculum consists primarily of six lessons, but it seems to me that these lessons can be summarized in one word: submission.

Robert Altmeyer's research shows that for a population of authoritarian submissives, authoritarian dominators are a survival necessity. Since those who learn their school lessons are too submissive to guide their own lives, our society is forced to throw huge wads of money at the rare intelligent authoritarian dominants it can find, from derivative startup founders to sociopathic Fortune 500 CEOs. However, with their attention placed on esteem, their concrete reasoning underdeveloped, and their school curriculum poorly absorbed, such leaders aren't well positioned to create value. They can create some, by imperfectly imitating established models, but can't build the abstract models needed to innovate seriously. For such innovations, we depend on the few self-actualizers we still get—people who aren't starving for esteem. And that does not include the wealthy, the powerful, and the "smart"; they learned their lessons well in school.

ARE WE BECOMING TOO CONNECTED?

GINO SEGRE
Professor of physics, emeritus, University of Pennsylvania; author, Ordinary Geniuses

Last year marked the twentieth anniversary of the World Wide Web coming into full existence. It was created at CERN, home of the world's largest particle accelerator, in response to the needs of large groups of experimenters scattered around the globe. They wanted a way to quickly and efficiently share data and analyses. WWW provided them with the tool. That model has been replicated over and over again. Our understanding of genetics, bolstered by international consortia of sequencers, is yet another example of this phenomenon.

The benefits of increased technological connectivity are so clear and appreciated that I need make no further effort to describe the gains. Rather let me consider what might be the negative effects of being so well connected.

I do so at the risk of sounding like one of those crotchety old guys whose every other pronouncement starts with *In my day, we used to do things differently.* Let me therefore make the disclaimer that I am not trying to vent my irritation at seeing the young texting their friends while I impart my so-called wisdom to them, or at the disconcerting effect of all too often being surrounded by individuals continually checking their iPhones for messages. Those are trivial annoyances and in any case have nothing to do with the argument at hand.

The issue of the threat of increased technological connectivity is not inconsequential. Potential losses that follow from it can be seen in the broader context of the lack of diversity following from homogenization of world culture and the dangers this poses for human evolution. But although sweeping syntheses of this sort can be drawn, I will limit myself to a few speculations regarding progress in science. In doing so, I almost entirely ask questions rather than, unfortunately, provide answers.

At a basic level, the threat shows up in academics in the matter of junior faculty appointments. A postdoctoral fellow now needs to publish frequently, and since the number of times he or she has been cited is part of the dossier, this requirement is hard to resist. But is it productive of thoughtful or even innovative science? In attempts to join the club, these efforts may be compromised. How likely are you to be promoted if you have not proceeded in lockstep along the golden path to more publications, more citations, and the all-too-precious grants? If you quit the mainstream and take the less trodden path, aren't you more likely to come to a career dead-end?

I am by no means advocating isolation, but it is overwhelming to see on my computer every morning the list, with synopses, of all the high-energy physics preprints submitted the day before. They appear complete with routing for access in PDF and alternative formats. Does this run the risk of producing a herd mentality? Submit now before you are scooped and it becomes too late! Can we imagine incidents when such pressure is unproductive? Does it encourage groupthink?

A comparison may be made to the way a budding scientist currently acquires information. The computer is the quick, easy, and efficient way of doing so, but the earlier stumbling through science journals in the library had its advantages. Though it led down many blind alleys, it was also a way of picking up and storing odd bits of information that might stimulate the wanderer in unforeseen ways.

Are we discouraging the oddball, the maverick, or the individual who wants to let a wild idea rumble around in the mind for a while? Let's not evoke yet again the image of Einstein toiling away unknown in a patent office but consider the less celebrated case of Max Delbrück. He was the son of a Berlin professor, earned his physics doctorate under Max Born, and was a postdoctoral fellow in Niels Bohr's Copenhagen Institute at a time when this was the straight and narrow path to success as a theoretical physicist. But at age twenty-six he opted to begin studying connections to biology. This eventually led to his investigating how viruses replicate. His first faculty appointment did not come until he was thirty-four. It was an instructorship at Vanderbilt University, hardly one's dream of a rapid rise through the ranks. But the work he had already done would earn him the Nobel Prize in physiology or medicine almost thirty years later.

It was never easy, but is it harder nowadays for a Delbrück to survive?

I believe the desire to make an unforeseen offbeat discovery is an integral part of what draws anyone to become a scientist and to persist in the quest. As is true in other walks of life, demands to conform intervene. It would be naïve to discount the struggle to obtain funds and the increasingly weighty burden that struggle imposes as research grows more expensive; that is yet another facet of the scientific life. But returning to my original message, isn't it possible that increased technological connectivity has subtle negative effects that should be considered as we praise the gains it offers us? As the march of science goes forward, perhaps we should heed a few warning signs along the road.

STRESS

ARIANNA HUFFINGTON

Chair, president, editor-in-chief, The Huffington Post Media Group; nationally syndicated columnist; author, Third World America

One of the things that worries me the most is the growing incidence of stress in our society. Over the last thirty years, self-reported stress levels have gone up 25 percent for men and 18 percent for women. Stress is a big contributor to the increase in diabetes, heart disease, and obesity. So it's a much bigger problem than most people realize, but thankfully there does seem to be a growing awareness of the destructive power and cost of stress—in terms of both dollars and lives. Stress wreaks havoc not just on our relationships, our careers, and our happiness but also on our health. On the collective level, the price we're paying is staggering—stress costs American businesses an estimated $300 billion a year, according to the World Health Organization. This is partly because stress was also the most common reason for long-term health-related absence, in a survey conducted by CIPD (Chartered Institute of Personnel and Development), the world's largest human resources association.

In the United States, 36 million adults suffer from high blood pressure that isn't being controlled, even though 32 million of them are receiving regular medical care. And nearly as many Americans—more than 25 million—have diabetes. The Centers for Disease Control and Prevention estimates that 75 percent of health-care spending is for chronic diseases that could be prevented. This is one reason why health-care costs are growing exponentially: Spending rose 3.8 percent in 2010 and 4.6 percent in 2011.

The easier, healthier, and cheaper way to treat stress is to deal with its causes instead of its effects. The good news is that we know what to do: Practices like mindfulness, meditation, yoga, and healthy sleep habits have proved extremely successful in combating stress. And awareness of the benefits of stress reduction is spreading, from the classrooms of the Harvard Business School, where students learn to better understand their emotions, to major corporations around the world, which have added meditation, mindfulness training, and yoga to the workplace. Olympic athletes have made napping and stress reduction part of their daily routine. Veterans and the medical professionals who treat them are increasingly embracing yoga as an effective way to navigate the consequences of PTSD.

Plus, new high-tech tools are making it possible for individuals to take more and more control of their own health. The first wave of connecting technology hyperconnected us to the entire world—but, in the process, often disconnected us from ourselves. That's why I'm so excited about the new wave of technology that reconnects us to ourselves. For example, a robust market of wearable devices has emerged, monitoring everything from activity and food intake to weight and sleep. Thirty million wearable devices were shipped in 2012, and 80 million are on track to ship in 2016—by which time consumer wearables will be a $6-billion market. These many innovations are signals of a significant change in attitude about the role technology can, and should, be playing in our lives.

PUTTING OUR ANXIETIES TO WORK

JOSEPH LeDOUX
Henry & Lucy Moses Professor of Science & professor of neuroscience & psychology, New York University; author, Synaptic Self: How Our Brains Become Who We Are

What should we be worried about? Pick your poison, or poisons. There's no shortage: Just read the front page of any major newspaper or watch the network and cable news shows. And if you are concerned that you haven't worried about the right things, or enough things, stop fretting. There are surely others who have it covered.

Ever since the phrase "the age of anxiety" entered the lexicon, each generation has claimed they have more to worry about than the previous one. But the fact is, anxiety is part of the human condition. It's the price we pay for having a brain that makes predictions—for the ability to see a future that is not necessarily foretold by the past.

Though we are an anxious species, we aren't all equally anxious. We each have our own set point of anxiety—a point toward which we gravitate. Ever notice how short-lived is the calm that results from eliminating a source of worry? Get rid of one and pretty soon something else takes its place, keeping each of us hovering around our special level of worry.

Anxiety can be debilitating. But even those who don't have an anxiety disorder still have stuff circling through their synapses that sometimes interferes with life's simplest chores. We don't necessarily want to get rid of anxiety altogether, as it serves a purpose: It allows us to focus our energy on the future. What we should worry about is finding some way to use, rather than be used by, our anxiety.

SCIENCE HAS NOT BROUGHT US CLOSER TO UNDERSTANDING CANCER

XENI JARDIN
Tech culture journalist; founding partner & coeditor, Boing Boing; *executive producer, host,* Boing Boing Video

We should be worried that science has not yet brought us closer to understanding cancer.

In December 1971, President Nixon signed the National Cancer Act, launching America's War on Cancer. Forty-odd years later, like the costly wars on drugs and terror, the War on Cancer has not been won.

According to the National Cancer Institute, about 227,000 women were diagnosed with breast cancer in the United States in 2012. And rates are rising. More women in America have died of breast cancer in the last two decades than the total number of Americans killed in World War I, World War II, the Korean War, and the Vietnam War combined.

But military metaphors are not appropriate to describe the experience of having, treating, or trying to cure the disease. Science isn't war. What will lead us to progress with cancer aren't better metaphors but better advances in science.

Why, more than forty years after this war was declared, has science not led us to a cure? Or to a clearer understanding of causes and prevention? Or simply to more effective and less horrific forms of treatment?

Even so, now is the best time ever to be diagnosed with cancer. Consider the progress made in breast cancer. A generation ago, women diagnosed with breast cancer would have had a prognosis entailing a much greater likelihood of an earlier death, more disfigurement, and a much lower quality of life during and after treatment.

Treatment-related side effects such as "chemobrain" are only just now being recognized as scientifically valid phenomena. A generation ago, breast cancer patients were told that the cognitive impairment they experienced during and after chemotherapy was "all in their heads," if you will.

Sure, there has been progress. But how much, really? The best that evidence-based medicine can offer for women in 2013 is still poison, cut, burn, then poison some more. A typical regimen for hormone-receptive breast cancer might be chemotherapy, mastectomy and reconstruction, radiation, at least five years of a daily anti-estrogen drug, and a few more little bonus surgeries for good measure.

There are still no guarantees in cancer treatment. The only certainties we may receive from our doctors are the kind no one wants. After hearing "We don't really know" from surgeons and oncologists countless times as they weigh treatment options, cancer patients eventually get the point. They really don't know.

We're still using the same brutal chemo drugs, the same barbaric surgeries, the same radiation blasts as our mothers and grandmothers endured decades ago—with no substantially greater ability to predict who will benefit and no cure in sight. The cancer authorities can't even agree on screening and diagnostic recommendations: Should women get annual mammograms starting at forty? Fifty? Or no mammograms at all? You've come a long way, baby.

Maybe, to get at the bottom of our worries, we should just "follow the money." Because the profit to be made in cancer is in producing cancer-treatment drugs, machines, surgery techniques—not in finding a cure or new ways to look at causation. There is likely no profit in figuring out the links to environmental causes: how what we eat or breathe as children may cause our cells to mutate; how exposure to radiation or man-made chemicals may affect our risk factors.

What can make you even more cynical is looking at how much money there is to be made in poisoning us. Do the dominant corporations in fast food, chemicals, agribusiness, want us to explore how their products affect cancer rates? Isn't it cheaper for them to simply pinkwash "for the cause" every October?

And for all the nauseating pink-ribbon feel-good charity hype (an industry in and of itself!), few breast cancer charities are focused on determining causation or funneling a substantial portion of donations to actual research and science innovation. Genome-focused research holds great promise, but funding for this science at our government labs, NIH and NCI, is harder than ever for scientists to secure. Why hasn't the Cancer Genome Atlas yielded more advances that can be translated now into more effective therapies?

Has the profit motive that drives our free-market society skewed our science? If we were to reboot the War on Cancer today, with all we now know, how and where would we begin?

The research and science that will cure cancer will not necessarily be done by big-name cancer hospitals or by Big Pharma. It requires a new way of thinking about illness, health, and science itself. We owe this to the millions of people who are living with cancer—or, more to the point, trying very hard not to die from it.

I know. I am one of them.

SOCIETY'S PARLOUS INABILITY TO REASON ABOUT UNCERTAINTY

AUBREY DE GREY
Gerontologist; cofounder & chief science officer, SENS Research Foundation; author, Ending Aging

Broadly well-educated people are generally expected by other broadly well-educated people to easily learn and accommodate new information on unfamiliar topics. By "accommodate," I mean absorption not only of facts but also of the general tenor of the topic as it exists in the expert community. Unfortunately, this expectation is frequently unfulfilled. I first became aware of the depth of this problem through my own work, the development of medical interventions against aging, in which the main problem is that throughout history we have had no choice but to put the horror of aging out of our minds by whatever psychological device, however irrational, may work—a phenomenon to which researchers in the field are, tragically, not immune (though that is changing at a gratifying pace). But here I wish to focus on a much more general problem.

Uncertainty is, above all, about time scales. Humans evolved in an environment where the short term mattered the most, but in recent history it has been important to depart from that mindset. What that means, in terms of ways to reason, is that we need to develop an evolutionarily unselected skill: how best to integrate the cumulative uncertainties that longer-term forecasting entails.

Consider automation. The step-by-step advance of the trend that began well before, but saw its greatest leap with, the Industrial Revolution has resulted in a seismic shift of work patterns from manufacturing and agriculture to the service industries. But, amazingly, there is virtually no appreciation of what the natural progression of this phenomenon—namely, the automation of service jobs, too—could mean for the future of work. What is left, once the service sector goes the same way? Only so many man-hours can realistically be occupied in the entertainment industry. Yet rather than plan for and design a world in which it is normal either to work for far fewer hours per week or for far fewer years per lifetime, societies across the world have acquiesced in a political status quo that assumes basically no change. Why the political inertia?

The main problem here is the public's deficiency in probabilistic reasoning. Continued progress in automation, as in other areas, certainly relies on advances that cannot be anticipated in detail and therefore not in precise time frames either. Thus it is a topic for speculation. I do not use that term in a pejorative way but to emphasize that aspects of the future about which we know little cannot thereby be ignored: We must work with what we have. And it is thought-leaders in the science and engineering realms who must take a lead here. Public policy overwhelmingly follows, rather than leads, public opinion: The number-one thing politicians work toward is getting reelected. Thus, while voters fail to reach objective conclusions about even the medium term—let's say a decade hence—it is fanciful to expect policy makers to do any better.

The situation is the worst in the extreme cases that can be summarized as "high risk, high gain"—low perceived probability of success but huge benefits in the event of success. As any academic will aver, the mainstream mechanisms for supplying public funding to research have sunk to a disastrous level of antipathy toward high-risk high-gain work, to the point where senior scientists stay one step ahead of the system by essentially applying for money to do work already largely complete and bearing no risk of not being delivered on time. The fields of research that most interest *Edge* readers are exceptionally susceptible to this challenge. Visionary topics are of necessity long-term, hence high risk, and of almost equal necessity high gain. In the area of medical research, for example: Are we benefiting the most people, to the greatest extent, with the highest probability, by the current distribution of research funding? In all such areas I can think of, the bias apparent in public opinion and public policy is in favor of approaches that might arguably (often very arguably) deliver modest short-term benefits but offer almost no prospect of more effective, second-generation approaches down the road. The routes to those second-generation approaches that show the best chance of success are marginalized because of their lack of "intermediate results."

We should be very worried about this. It is already costing masses of lives, by slowing down life-saving research. And how hard is it to address, really? How hard is Bayes' Theorem, really? The single most significant thing that those who understand this issue can do to benefit humanity is agitate for better understanding of probabilistic reasoning among policy makers, opinion formers, and thence the public.

THE RISE IN GENOMIC INSTABILITY

ERIC J. TOPOL, M.D.
Professor of genomics, Scripps Research Institute; director, Scripps Translational Science Institute; author, The Creative Destruction of Medicine

We were taught wrong. The old elementary-school science lesson that our DNA sequence is a fifty-fifty split of our mother and father omitted the salient matter that there are new, so-called *de novo* mutations that spontaneously occur and are a big deal.

It wasn't until this past year that we could sequence whole human genomes of families, and even of single sperm cells, to directly quantify how frequently these *de novo* mutations arise. Each of us has between about eighty and a hundred changes in our native (germline) DNA that are not found in our parents' DNA. But the source is the genomic instability from their eggs or sperm. On average, about fifteen to twenty "spelling errors" come from our mother and thirty to sixty from our father. We have a new appreciation of the father's biologic clock, with aging dads having more sperm DNA instability, and increasing evidence that this phenomenon is linked to a higher risk of autism and schizophrenia.

Although the new mutations are rare in any given individual—representing a tiny fraction, less than 0.001 percent of your genome—the chance that they will do harm is great. That's because they are not subject to natural evolutionary selection. Whereas there's a small chance the mutation could have a positive effect, the overwhelming likelihood is for a deleterious one, as we've seen with the recent studies of such mutations in children with severe intellectual disability and other neurodevelopmental diseases.

So, to respond to the 2013 *Edge* Question, What's the worry? We should be concerned that this genomic instability in our germline DNA, and also in our somatic (body cell) DNA, is on the rise. We're seeing more new cases of cancer, which represent DNA off the tracks, prototypic of genomic instability. And while the aging-father trend is a clear and global phenomenon and may contribute to a small part of the increased incidence of autism, the story may be much bigger than that. So far, we know only that there exists a relationship with easy-to-diagnose traits like schizophrenia and severe cognitive disability. What about the more subtle effect of such mutations on other conditions, such as mild cognitive impairment or susceptibility to diabetes? The question we really need to ponder is precisely why genomic instability is increasing with age, and why more people are getting diagnosed with cancer from year to year—about 1.7 million Americans in 2012 and an increased incidence for seven of the leading types of cancer, even adjusted for the advancing age of the population.

I think a significant portion of genomic instability is due to environmental effects. For example, exposure to increased radiation is a prime suspect—be it man-made thermal radiation from atmospheric greenhouse gases or via medical imaging that uses ionized radiation. There are probably many other environmental triggers in our "exposome" that have yet to be unraveled, such as the interaction of our native DNA with our gut microbiome or the overwhelming pervasive exposure we have to viruses that can potentiate genomic instability.

Although we now have an appreciation for the frequency of *de novo* mutations due to the spectacular advances in sequencing technology and analytics, we don't have even a rudimentary understanding of what induces them in the first place—or of the more subtle effects that may track with a sort of devolution of humankind. What is especially disconcerting is that the signals of increased genomic instability are occurring in a relatively short time span in the context of human evolution over millions of years.

Thus it may take a long time for this to play out. But we could do something about it now, by conducting an in-depth study of *de novo* mutations and environmental interactions among hundreds of thousands of individuals and their offspring. Not that we want to suppress all human *de novo* mutations, but perhaps someday there will be a way forward to prevent or screen out deleterious ones—and foster those that prove favorable. It might be considered by some to represent unnatural selection, but that's what it may take to turn the tide.

CURRENT SEQUENCING STRATEGIES IGNORE THE ROLE OF MICROORGANISMS IN CANCER

AZRA RAZA, M.D.

Professor of medicine; director, the MDS (Myelodysplastic Syndrome) Center, Columbia University

As a researcher studying cancer for almost four decades, I have witnessed several cycles during which the focus of investigators has shifted radically to accommodate the prevailing technical or intellectual advances of the time.

In the 1970s, it was discovered that while the use of single chemotherapeutic drugs produced impressive results in certain cancers, adding more agents could effectively double the response rate. Thus the seventies were dedicated to combination chemotherapies. The eighties were dominated by a race to identify mutations in the human homologs of genes that cause cancers in animals (oncogenes).

This was followed in the 1990s by a focus on immune therapies and monoclonal antibodies, resulting in some resounding successes in the treatment of lymphomas. Given the technical advances as a result of the Human Genome Project, the spotlight in this decade has now swung toward developing the Cancer Genome Atlas, utilizing high-throughput genome analysis to catalog genetic mutations in some of the most common cancers.

The premise here is that by identifying mutations in cancer cells and comparing them to normal cells of the same individual, a better understanding of the malignant process, and new targets for treatment, will emerge. This is all very exciting, but if the current trend of sequencing the cancer genome continues unchanged, the role of pathogens in initiating and/or perpetuating cancer may be missed for a long time to come. Here's why.

Chances of getting cancer in a lifetime are 1 in 2 for men and 1 in 3 for women. The most widely accepted view is that cancer is a genetic disease. While there's no doubt that cancer is caused by mutated cells, the question is, How do the mutations arise in these cells in the first place? Among several possibilities, there are at least two that could relate the presence of a cancerous growth to pathogens.

The first is that the individual is born with a mutated cell but the cell remains dormant because the microenvironment (soil) where it resides may not be suitable for its proliferation. If the surrounding conditions change in such a way that the soil is now more fertile for the abnormal cell at the expense of normal cells, then the mutated cell could expand its clonal population, resulting in a cancerous growth. It is a well-appreciated fact that most cancers thrive in a pro-inflammatory microenvironment, and pathogens are capable of altering a normal microenvironment to a pro-inflammatory one, thus providing the required conditions for a mutated cell to grow.

A second possibility is that the pathogen infects a normal cell and co-opts its vital machinery, resulting in unchecked growth of the mutated cells. Thus it is not unreasonable to look for the presence of a microorganism in either the malignant cell or in its microenvironment.

Studies have already demonstrated that 20 to 30 percent of cancers worldwide are associated with chronic infection: The Epstein–Barr virus can cause lymphomas and nasopharyngeal cancers; the human papillomavirus (HPV) causes cervical cancer and head and neck tumors; long-term bacterial infection with *Helicobacter pylori* can cause stomach cancer; and hepatitis B virus (HBV) causes liver cancer. There is a good possibility that many more cancers will be associated with pathogens, and some cancer-causing pathogens may turn out to be part of our collective microbiome, the community of microorganisms that live in our bodies. The microbiome makes up about 1 to 3 percent of our biomass and outnumbers human cells in the body by 10 to 1.

The Human Microbiome Project (HMP), a consortium of eighty universities and scientific institutions funded by the National Institutes of Health, is looking at this aspect tangentially, through defining the communities of bacteria that live within the human body and their role in health and disease.

Unfortunately, HMP is only defining the communities of bacteria (and they have shown 10,000 species!) that live in our bodies, but this will not help identify microorganisms associated with cancers. The reason that more cancers have not been associated with microorganisms thus far is because we lack suitable techniques for detecting devious pathogens like retroviruses. Unfortunately, only a small proportion of effort and money is being invested in this area of research.

Most researchers are studying malignant diseases through sequencing, but microorganisms associated with cancer will be missed, because genomic sequencing, which should involve whole-genome sequencing (100 percent of DNA), has been mostly abbreviated to whole-exome sequencing (the gene-coding region, comprising about 2 percent of DNA). This has happened at least in part because massive parallel sequencing is still not economically feasible or technically advanced enough for routine use. The problem is that the integration of a cancer-causing pathogen is not likely to be in the coding region (exome), which is now the main focus of study for the investigators who continue to be the principal drivers of cancer research.

The second problem is that only the malignant cells are being studied by the vast majority of researchers, whereas the cancer-causing microorganism may be residing in the cells of the tumor microenvironment, rendering the soil fertile only for the growth and expansion of the mutated cell. In order to develop a more comprehensive understanding of the cancerous process, it is important to study both the seed and the soil and to perform whole-genome sequencing rather than examining only the coding regions of the genome.

Carl Zimmer writes in his book *Parasite Rex* that parasites "are expert at causing only the harm that's necessary, because evolution has taught them that pointless harm will ultimately harm themselves." Evolution will eventually teach the malignant cells (and their masters) that the only way they will succeed in perpetuating themselves is by immortalizing rather than killing the host. Let us hope we will have the answer to the cancer debacle before that.

THE FAILURE OF GENOMICS FOR MENTAL DISORDERS

TERRENCE J. SEJNOWSKI

Computational neuroscientist, Francis Crick Professor, Salk Institute for Biological Studies; coauthor (with Patricia Churchland), The Computational Brain

The Human Genome Project was a great success. Not only do we have the human genome, but after a decade of advances in gene sequencing the cost has plummeted from a few billion dollars per genome to a few thousand. This has had a profound effect on what questions we can ask about biological systems and has generated many important results.

Mental disorders are major health problems for society. Most of us know someone with an autistic child, a schizophrenic cousin, a friend suffering from depression. What impact has genomics had on the treatment of these disorders?

Autism occurs in 1 percent of children under the age of eight and their care is estimated to cost society $3.2 million over a lifetime; the annual cost for all autistic people in the United States is $35 billion per year. Schizophrenia, whose symptoms first appear in early adulthood, affects 1 percent of the population—annual cost, $33 billion. In comparison, the average annual cost of the war in Afghanistan has been $100 billion. We are waging multiple wars against mental disorders, with no end in sight. The burden on families and caregivers is not just monetary: Each person with a major mental disorder can disrupt many other lives.

Both autism and schizophrenia have substantial inherited components, and there was great hope that the origin of mental disorders could be understood by identifying the genes responsible. In studies of monozygotic twins, the concordance for autism is 30 to 90 percent—40 to 60 percent for schizophrenia. Large-scale genome-wide association studies have screened thousands of families with these disorders and concluded that no single gene mutation, insertion, deletion, or copy number variation can account for more than a small fraction of the variance in the population. These studies cost hundreds of millions of dollars and have lists of authors as long as those on the Higgs boson discovery paper. Hundreds of genes have been implicated, many of them known to be important for synapse development and function. Because autism and schizophrenia are far from being Mendelian traits, it is much more difficult to identify therapeutic targets that would be effective for a wide range of patients. This was a great disappointment and a concern for future genomics research on mental disorders.

Although sequencing the human genomes of patients has not yielded direct benefits, genetic tools are nonetheless opening up new approaches for treating mental disorders. Clinical depression is another debilitating disorder, affecting 15 million Americans; 20 percent of them do not respond to antidepressants. The annual cost for depression is $83 billion. A promising new therapy for drug-resistant depression is electrical stimulation of the anterior cingulate cortex, which is connected with other brain regions important for regulating well-being, whether we feel safe or vulnerable, and especially our emotional responses. In some cases, effects are dramatic, with the veil of depression lifting minutes after the onset of electrical stimulation. Although deep brain stimulation is promising, we don't know why it works; progress will depend on more precise control of neural activity. A new technique that could revolutionize the treatment of depression and other brain disorders, such as Parkinson's disease, is based on stimulating neurons with light rather than microelectrodes. Optogenetics allows light-stimulated ion channels to be selectively delivered to neurons via viruses. Depending on the ion channel that is inserted, light can cause a neuron to spike or be silenced.

Psychosis is currently treated with drugs that are at best palliative, with debilitating side effects. Progress in improving the treatment of mental disorders has been slow, but there are reasons to be optimistic. That depression can be lifted so quickly suggests that the neural circuits are intact but in an imbalanced state. Gross electrical stimulation may compensate in ways not yet understood. In schizophrenia, there is evidence for an imbalance between the excitation and inhibition in cortical circuits; in particular, there is downregulation of GABA, an inhibitory neurotransmitter, in an important class of inhibitory interneurons that provides negative feedback to the excitatory pyramidal neurons in the cerebral cortex. Plans to record from a million neurons simultaneously using nanotechnology are under way; this will give us a much better map of brain activity in normal and abnormal states. As we learn more about the nature of these imbalances, and as molecular techniques for manipulating neural circuits are perfected, it may be possible to better treat the symptoms of major mental disorders and perhaps even cure them.

There is a wide spectrum of symptoms and severity among those diagnosed with autism and schizophrenia. We now know that this depends in part on the particular combination of genes affected. Environmental factors also have a major influence. To paraphrase Tolstoy: Happy brains are all alike; every unhappy brain is unhappy in its own way.

EXAGGERATED EXPECTATIONS

STUART FIRESTEIN
Professor & chairman, Department of Biological Sciences, Columbia University; author, Ignorance: How it Drives Science.

How often does science fail to deliver? How often should it fail? Should we be worried about the failure rate of science?

Much has been made recently of all the things that science has predicted which haven't come true. This is typically presented as an indictment of science and sometimes even a reason not to put so much faith in it. Both of these are wrong. They are wrong on the statistics and they are wrong on the interpretation.

The statistical error arises from lumping together scientific predictions that are different in kind. For example, the covers of postwar popular-science and technology magazines are full of predictions of amazing developments that were supposed to be just around the corner—floating airports, underground cities, flying cars, multilayer roadways through downtowns, et cetera, ad nauseam. Very few of these wild predictions came to pass, but what was the chance they would? They were simply the unbridled imaginings of popular-science writers or graphic artists looking for a dramatic cover that would sell magazines.

You can't lump these predictions in the same bin with more serious promises, like the eradication of cancer. One sort of prediction is just imaginative musing; the other is a kind of promissory note. Artificial intelligence, space travel, alternative energies, and cheaper, more plentiful food are all in this second category. They cost money, time, and resources, and they are serious activities; the speculations of science-fiction writers have no cost associated with them.

But of course not all the serious promises have worked out, either. The second error, one of interpretation, is that we have therefore squandered vast sums of public money and resources on abject failures. To take one case, consider the so-called War on Cancer. We have spent $125 billion on cancer research since President Richard Nixon "declared war" on this disease forty-two years ago, in 1971. The result: Over the same forty-two-year period, some 16 million people have died from cancer and it is now the leading cause of death in the U.S. Sounds bad, but in fact we have cured many previously fatal cancers and prevented an unknowable number of cases by understanding the importance of environmental factors (asbestos, smoking, sunshine, etc.).

And what about all the ancillary benefits that weren't in the original prediction—vaccines, improved drug-delivery methods, sophisticated understanding of cell development and aging, new methods in experimental genetics, discoveries that tell us how genes are regulated (all genes, not just cancer genes), and a host of other goodies that never get counted as resulting from the War on Cancer. Then there is the unparalleled increase in our understanding of biology at every level—from biochemical cascades to cell behavior to regulatory systems to whole animals and people, and the abovementioned effects of the environment on health. Is anybody tallying all this up? I'd venture to say that, all in all, this cancer war has given us more for the dollars spent than any real war, certainly any recent war.

Much of science is failure, but it is a productive failure. This is a crucial distinction in how we think about failure. More important is that not all wrong science is bad science. As with the exaggerated expectations of scientific progress, expectations about the validity of scientific results have simply become overblown. Scientific "facts" are all provisional, all needing revision or sometimes even outright upending. But this is not bad; indeed, it is critical to continued progress. Granted it's difficult, because you can't just believe everything you read. But let's grow up and recognize that undeniable fact of life—not only in the newspapers but in scientific journals.

In the field of pharmacology, where drugs are made, we say that the First Law of Pharmacology is that every drug has at least two effects—the one you know and the other one (or ones?). And the even more important Second Law of Pharmacology is that the specificity of a drug is inversely proportional to the length of time it is on the market. This means, in simpler terms, that when a drug first comes out, it is prescribed for a specific effect and seems to work well. But as time passes and it is prescribed more widely and taken by a more diverse population, then side effects begin showing up; the drug is not as specific to the particular pathology as was thought, and it seems to have other unexpected effects, mostly negative. This is a natural process. It's how we learn about the limitations of our findings. Would it be better if we could shortcut this process? Yes. Is it likely we'll be able to? No—unless we would be content with trying only very conservative approaches to curing diseases.

So what's the worry? That we will become irrationally impatient with science, with its wrong turns and occasional blind alleys, with its temporary results that need constant revision. And we will lose our trust and belief in science as the single best way to understand the physical universe (which includes us, or much of us). From a historical perspective, the path to discovery may seem clear, but the reality is that there are twists and turns and reversals and failures and cul-de-sacs all along the path to any discovery. Facts are not immutable, and discoveries are provisional. This is the messy process of science. We should worry that our unrealistic expectations will destroy this amazing mess.

LOSING OUR HANDS

SUSAN BLACKMORE
Psychologist; author, Zen and the Art of Consciousness

I don't mean that someone is going to come and chop our hands off. I mean that we are unwittingly but eagerly outsourcing more and more of our manual skills to machines. Our minds are losing touch with our bodies and the world around us and being absorbed into the evolving technosphere.

To begin with, we created machines to do our bidding and make our lives easier and more enjoyable, but we have failed to notice how quickly that relationship is changing. What began as master and servant is heading for "obligate symbiosis," a state in which neither can survive without the other.

These machines include everything from the high-powered physical machines that build our roads or harvest our crops to the thinking machines in everything from toasters to Internet servers. Engines and cranes very obviously relieve us of heavy manual labor, but in the process they also change the nature of our minds. This is because manual skills are really not just about hands; they are about the way our brains and hands interact. When I learn how to plant potatoes, turn a chair leg, or replace a roof tile, I am not just learning intellectually how far apart to space the tubers, or the principles of using a lathe; I am involving my whole body and mind in learning a new skill. This takes time and practice. It changes me gradually into someone who will more easily learn how to grow beans, carve a chair seat, or mend the gutters.

We can see the loss of these skills in the obvious fact that fewer people now learn them. How many of us could build a waterproof shelter, make furniture, or even grow our own food? These memes still survive, especially in less developed cultures, but the numbers are falling. Just as worrisome is changing attitudes. For example, in the British education system, crafts like woodworking and cooking or trade skills like bricklaying and plumbing are now tested more by written exams than by what students can actually do. This is meant to raise the status of these subjects, but instead it turns them into purely intellectual knowledge, belittling the important manual skills that take so much practice to acquire. Every time we build a machine to do something we previously did ourselves, we separate our minds a little further from our hands.

Perhaps less obvious is that the same process is going on as we enthusiastically adopt communications technology. When we began using e-mail, it seemed a handy replacement for the slow process of sending letters. When we got our first mobile phone, it seemed just a more convenient way of talking to people. But look at smartphones now. No one can compete in today's world without using at least some of this technology. Opting out to become "self-sufficient" is even more hopeless than it was in the 1970s, when many of us flirted with the idea.

Yet somehow we cling to the notion that because we invented these machines in the first place, they are still there for our benefit and we can do what we like with them. This is obviously not true. From a meme's-eye perspective, it is the technomemes and the wonderful machinery that copies, recombines, stores, and propagates them that benefit—not us. It is they who are rapidly evolving, while our bodies hardly change at all.

But the way we use them is changing. Our hands now spend little time making or growing things and a lot of time pressing keys and touching screens. Our brains have hardly changed in size or gross structure, but their function has. Our evolved desires for fun, competition, and communication lead us into ever vaster realms of online information and away from the people right next to us. And who are "we"? Our selves, too, are changing as they disconnect from our bodies, becoming as much the person who exists on multiple Web sites and forums as the physical body who acts and interacts right here and now—as much a digitally propagated entity as the man now holding my hand in his.

So what should worry us now is our role in this world. If we are not masters in control of our technology, who or what are we becoming?

Here is a possible analogy. About 2 billion years ago, mitochondria evolved from primitive bacteria by entering into a symbiotic relationship with early eukaryotic cells. Each benefited the other, so this was not a hostile takeover but a gradual coming together until neither the living cells nor the mitochondria within them could survive without the other. The cells feed and protect the mitochondria; the mitochondria provide the power. Could our future be heading in that direction? The analogy implies a world in which humans manage the power supplies to feed an ever increasing number of inventions, in return for more fun, games, information, and communications—a world in which we so value the fruits of our machines that we willingly merge both physically and mentally with them.

The prospect looks bleak. The demands of this evolving system are insatiable, and the planet's resources are finite. Our own greed is insatiable, and yet its satisfaction does not make us happier. And what if the whole system collapses? Whether it's climate change, pandemics, or any of the other disaster scenarios we worry about, there might indeed come a time when the banks collapse, the power grids fail, and we can no longer sustain our phones, satellites, and Internet servers. What then? Could we turn our key-pressing, screen-swiping hands to feeding ourselves? I don't think so.

What should worry us is that we seem to be worrying more about the possible disasters that might befall us than who we are becoming right now.

LOSING TOUCH

CHRISTINE FINN

Archaeologist, journalist; author, Artifacts: An Archaeologist's Year in Silicon Valley

My worry? Losing touch.

Biologists at the University of Newcastle, U.K., recently published a report claiming that fingers wrinkling in a long bath was a sign of evolutionary advantage. The prunelike transformation provided the digits with a better grip. More than helping one to grasp the soap, this skin puckering suggested itself as a factor for survival, as our Mesolithic ancestors foraged for food in rivers and rock pools.

That widely reported reminder of our evolved capabilities helped assuage my own real worry—call it, perhaps, a haptical terror—about losing touch with the physical world. What is the future for fingers, as tools, in the Digital Age? Now that the latest interface is a touch that is smooth and feather-light, and human-to-machine commands are coming to be spoken, or breathed, or blinked, or even transmitted by brain waves, will finger-work be the preserve solely of artists and child's-play? Fingers could still form churches and steeples and all the peoples, be in the play of poets, peel an orange. But in the Digital Age, will there be pages still to turn, tendrils to be untangled, a place for hard keystrokes, not simply passing swipes? After all, the digit, birthmarked with its unique code, is our security guard, a hush as much as the finger placed to the mouth.

But the fingers are fighting back.

We are encouraged to wield kitchen tools, find grandma's sewing box. To beat eggs, pound dough, and ice cakes; to join knitting circles; to plunge our hands into landscape to pick wild food. Things are still palpable; real book sales are encouraging, and we are still hooked on marginalia and turning down the corner of a page.

A few months ago, I visited the Florida house where Jack Kerouac wrote *Dharma Bums* in an ecstatic burst of typing over eleven days and nights. In this refuge from New York critics of *On the Road*, his fingers translated brain to hand to brain, synapses sizzling from caffeine and Benzadrine. In these wooden rooms nearly sixty years later could be heard the distinctive sound of a metal typewriter. Not Jack's ghost pounding out but the latest writer-in-residence, a young woman in her twenties who, it transpired, had arrived from Ohio with both a laptop and a vintage manual typewriter, to know what it felt like.

This essay began with the proverbial note sketched out by hand on the back of an envelope. My fingers then picked out the words on my smartphone. My worry had been that these two processes, necessary to shape and synthesize, were somehow conflicted. I worried about our grip on technology. But the wrinkling of our fingers reminds me that we engage with the evolved and the still evolving.

THE HUMAN/NATURE DIVIDE

SCOTT SAMPSON
Dinosaur paleontologist; research curator, Utah Museum of Natural History; author, Dinosaur Odyssey: Fossil Threads in the Web of Life

We should all be worried about the gaping psychological chasm separating humanity from nature. Indeed, a strong argument can be made that bridging this divide deserves to be ranked among the most urgent 21st-century priorities. Yet so far the human/nature divide hasn't even made it to our cultural to-do list.

For the past several decades, numerous scientists and environmentalists have been telling us we must change our ways and strike a balance with nature or face catastrophic consequences. I have often participated in this echo chamber, doling out dire statistics in hopes of engaging people in action. The unspoken assumption has been that cold, hard facts are all that's needed for people (including businesspeople and elected officials) to "get it" and alter their unsustainable ways. To date, however, virtually all the key indicators—from greenhouse-gas emissions to habitat and species losses—are still heading in the wrong direction.

The problem is that humans aren't rational creatures. At least not when it comes to shifting their behaviors. As marketing executives have long understood, humans are far more susceptible to emotional messages, especially when conveyed through imagery. Want to escalate sales of some new car model? Beautiful people driving through pristine natural settings are far more powerful motivators than statistics on horsepower and fuel efficiency.

But what emotion is missing? What emotion do we need to foster a sustainable shift in human behavior? In a word, love.

As the late evolutionary biologist Stephen Jay Gould once claimed in an uncharacteristic moment of sentimentality, "We cannot win this battle to save species and environments without forging an emotional bond between ourselves and nature as well—for we will not fight to save what we do not love." The good news is that thanks to a lengthy evolutionary tenure living in intimate contact with the nonhuman world, the capacity for forming an emotional attachment with nature probably lies dormant within all of us, waiting to be reawakened (think E. O. Wilson's "biophilia").

The bad news is that as a species, we've never been more disconnected from the natural world. Thanks to a variety of factors—among them fear of strangers and an obsession with screens—children's firsthand encounters with nature in the developed world have dropped precipitously, to less than 10 percent of what they were just one generation ago. The average American youth now spends seven to ten hours per day staring at screens, compared to a mere handful of minutes in any "natural" setting. The result of this indoor migration is a runaway health crisis, both for children (obesity, ADHD, stress, etc.) and the places where they live.

Science has been one of the primary forces driving a wedge between humans and nature, prompting us to see nature as objects rather than subjects, resources to be exploited rather than relatives to be respected. Yet science, particularly over the past few decades, has also empirically demonstrated our complete embeddedness within nature, from the trillions of bacterial cells that far outnumber human cells in our bodies to our role as newbie actors in the 14-billion-year evolutionary epic.

Do we need more science? Of course, and the general public must learn the necessary facts, dire and difficult though they may be. We're also going to need all the technological help we can get in navigating a sustainable path into the future. Yet knowledge and technology without emotional connection simply won't cut it. The next generation of humans must learn to see their relationship with the natural world in ways that will seem alien to our current anthropocentric, reductionist, and materialistic perspective.

POWER AND THE INTERNET

BRUCE SCHNEIER
Security technologist; author, Liars and Outliers: Enabling the Trust That Society Needs to Thrive

All disruptive technologies upset traditional power balances, and the Internet is no exception. The standard story is that it empowers the powerless, but that's only half the story. The Internet empowers everyone. Powerful institutions might be slow to make use of that new power, but since they are powerful they can use it more effectively. Governments and corporations have woken up to the fact that not only can they use the Internet but they can also control it for their interests. Unless we start deliberately debating the future we want to live in, and the role of information technology in enabling that world, we will end up with an Internet that benefits existing power structures and not society in general.

We've all lived through the Internet's disruptive history. Entire industries, like travel agencies and video rental stores, disappeared. Traditional publishing—books, newspapers, encyclopedias, music—lost power, while Amazon and others gained. Advertising-based companies like Google and Facebook gained a lot of power. Microsoft lost power (as hard as that is to believe).

The Internet changed political power as well. Some governments lost power as citizens organized online. Political movements became easier, helping to topple governments. The Obama campaign made revolutionary use of the Internet, both in 2008 and 2012.

And the Internet changed social power, as we collected hundreds of "friends" on Facebook, tweeted our way to fame, and found communities for the most obscure hobbies and interests.

And some crimes became easier: Impersonation fraud became identity theft, copyright violation became file sharing, and accessing censored materials—political, sexual, cultural—became trivially easy.

Now powerful interests are looking to deliberately steer this influence to their advantage. Some corporations are creating Internet environments that maximize their profitability—Facebook and Google among many others. Some industries are lobbying for laws that make their particular business models more profitable: Telecom carriers want to be able to discriminate between different types of Internet traffic, entertainment companies want to crack down on file sharing, advertisers want unfettered access to data about our habits and preferences.

On the government side, more countries censor the Internet—and do so more effectively—than ever before. Police forces around the world are using Internet data for surveillance, with less judicial oversight and sometimes in advance of any crime. Militaries are fomenting a cyberwar arms race. Internet surveillance—both governmental and commercial—is on the rise, not just in totalitarian states but in Western democracies as well. Both companies and governments rely more on propaganda to create false impressions of public opinion.

In 1996, cyberlibertarian John Perry Barlow issued his Declaration of the Independence of Cyberspace. He told governments: "You have no moral right to rule us, nor do you possess any methods of enforcement that we have true reason to fear." It was a utopian ideal, and many of us believed him. We believed that the Internet generation, those quick to embrace the social changes this new technology brought, would swiftly outmaneuver the more ponderous institutions of the previous era.

Reality turned out to be much more complicated. What we

forgot is that technology magnifies power in both directions. When the powerless found the Internet, suddenly they had power. But while the unorganized and nimble were the first to make use of the new technologies, eventually the powerful behemoths woke up to the potential—and they have more power to magnify. And not only does the Internet change power balances but the powerful can also change the Internet. Remember how incompetent the FBI was at investigating Internet crimes in the early 1990s? Or how Internet users ran rings around China's censors and Middle Eastern secret police? Or how digital cash was going to make government currencies obsolete, and Internet organizing was going to make political parties obsolete? Now all that feels like ancient history.

It's not all one-sided. The masses can occasionally organize around a specific issue—SOPA/PIPA, the Arab Spring, and so on—and can block some actions by the powerful. But it doesn't last. The unorganized go back to being unorganized, and powerful interests take back the reins.

Debates over the future of the Internet are morally and politically complex. How do we balance personal privacy against what law enforcement needs to prevent copyright violations—or child pornography? Is it acceptable to be judged by invisible computer algorithms when being served search results? When being served news articles? When being selected for additional scrutiny by airport security? Do we have a right to correct data about us? To delete it? Do we want computer systems that forget things after some number of years? These are complicated issues that require meaningful debate, international cooperation, and iterative solutions. Does anyone believe we're up to the task?

We're not, and that's the worry. Because if we're not trying to understand how to shape the Internet so that its good effects outweigh the bad, powerful interests will do all the shaping. The Internet's design isn't fixed by natural laws. Its history is a fortuitous accident: an initial lack of commercial interests, governmental benign neglect, military requirements for survivability and resilience, and the natural inclination of computer engineers to build open systems that work simply and easily. This mix of forces that created yesterday's Internet will not be trusted to create tomorrow's. Battles over the future of the Internet are going on right now: in legislatures around the world, in international organizations like the International Telecommunication Union and the World Trade Organization, and in Internet standards bodies. The Internet is what we make it, and is constantly being re-created by organizations, companies, and countries with specific interests and agendas. Either we fight for a seat at the table or the future of the Internet becomes something that is done to us.

CLOSE TO THE *EDGE*

KAI KRAUSE

Software pioneer; software and graphical user interface designer; philosopher

We all *should* be worried . . . that somewhere in New York there is a powerful cultural entrepreneur who is surrounding himself with a couple hundred of the smartest brains and deploying his army of gray matter on . . . *coming up with more stuff to worry about.*

Like we have run out of daily doses of crises and cliffs and badly need fresh doom to go with the gloom?

Really, I had hoped that we could have turned this exercise in a useful and positive direction. For instance: *"Make a concrete proposal for how $1 billion could be put to the most effective use in 2013"*—and then see what that brain pool is really capable of. *And actually do it.*

There are a few billionaires and many philanthropists among the group. One could quite easily set up an *Edge* Prize to rival the X Prize, posing concrete problems to overcome in the smartest manner.

This could include science problems per se but also the meta problem of *science itself* being undervalued, and indeed orphaned to some degree. The Higgs boson means very little to the hick morons who could do with increased education at all levels. And more than bringing up the bottom, there is enormous dormant potential at the top—the smart ones in every class in Anytown, anywhere.

Just as an example: Imagine for one second if literally every child were issued its own xPad. *Entirely free.* All of them with identical specs, totally shatterproof. With a cloud infrastructure to provide a home for each student, a thin front end containing hardly anything locally, each device just a porthole, neutral and exchangeable. There would be no point in stealing one, a valueless commodity.

The sheer size of the target audience would allow third parties to create brilliant courseware. It would equalize the playing field and raise awareness at all levels. New kinds of classes could be offered, such as how to find answers and solutions, how to search, how to deal with security, the pleasures and pitfalls of the social networks—all those issues that are of much higher importance to kids than the "old-school" curriculum—and which they are left to deal with on their own now.

There are about 80 million people enrolled in education in the United States. Total budget is almost a trillion. You think such a move could make a difference? You bet it would. An extremely cost-effective one, too. There are endless *but but buts* to intervene, sure. Lots of detail questions, lots to think about. But *that* is where a think tank of smart folks should be put to great use.

Just making that itself an *Edge* Prize: the best proposals for implementation, leading to hundreds of them, with a system for peer review to *let the truly smart ideas bubble to the top*—as opposed to handing it off to some supposed "expert" study group, glacial government processes, or lobbied interests of the industry. You ask this question in the schools themselves, and you would be amazed at the number of detailed problems you can worry about . . . but also solutions and suggestions!

The trick would be *to create the system* in which the best ideas can be rewarded. You really think Apple needs to have $126 billion in cash reserves and would not be fine with $124 billion? I bet they would love to help, if not for entirely altruistic reasons—and so would all the others. Given the proper proposal, at a large scope, it is actually not the money that is the problem.

And there you are, with what I feel. We *should* be worried *that this is not happening.*

And I do not mean this one lone xPad example idea, pulled out of a hat. I mean the principle itself: We don't see that all the problems have become much too complex to be solved by our current methodology: The way the government is defined, the way it is run, the way it is chosen, the way it is funded, how it deploys the funds—*it is all broken. . . .* Nothing to do with left or right, but seeing it as abstractly as possible, *the individuals, in their contract with society as a whole*, are on the edge of canceling.

And the only way out of all this is applied sciences, if you think about it: *Smart thinking, intelligent planning, systematic analysis;* beyond partisan opinions, outside corporate brands, without financial gain. Dealing with it almost as an art form: *the beauty of an optimal path, the pleasure of finding a solution.*

And there *are* the people who can do it, who have dedicated themselves to exactly these processes all their lives—alas, only in their narrow sub-sub-sub-fields.

The state of the planet is the crowning super problem of them all—and yet the people who may be able to provide insights and offer fresh ideas are often occupied by trivial and mundane side issues, it seems. Such as answering questions about whether they can think of more things to worry about.

THE PARADOX OF MATERIAL PROGRESS

ROLF DOBELLI
Founder, Zurich.Minds; journalist; author, The Art of Thinking Clearly

I recently had dinner with a friend, a prominent IP lawyer, at his mansion in Switzerland, one of the few spots directly on Lake Zurich. As is customary with people who have mansions, he gave me the complete tour, not leaving out the sauna (how many different ways are there to decorate a sauna?). The mansion was a fireworks display of technological progress. My friend could regulate every aspect of every room by touching his iPad. "Material progress," he said during his show, "will soon come to every home." Stories of high-tech, high-touch houses have been around for decades, but it was still neat to see that it finally exists. Sensing my lack of amazement, he guided me to his "picture room." Photographs on display showed him with his family, on sailboats, on ski slopes, golf courses, tennis courts, and horseback. One photo he seemed especially proud of showed him with Pope Benedict XVI. "A private audience," he said.

So what do we learn from this that we didn't learn from *The Great Gatsby*?

Material progress will continue to spread. Knowledge is cumulative. At times in our past, knowledge has diminished. The classic case is Tasmania, or—on a grander scale—the Middle Ages. But since Gutenberg it is difficult to imagine that humanity will ever again shed information. Through the accumulation of knowledge and global trade, the goods and services that my lawyer friend enjoys today soon will be available to the poorest farmer in Zimbabwe. But no matter how much knowledge we accumulate, no matter how cheap computation, communication, and information storage become, no matter how seamlessly trade flows, that farmer will never get any closer to a date with the Pope.

See the Pope allegorically as all the goods and services that are immune to technological creation and reproduction. You can vacation on only one St. Barts. Rauschenberg created just a few originals. Only so many mansions dot the lakeshore in Zurich. Bringing technology to bear won't help create any more. A date with a virtual Pope will never do the trick.

As mammals, we are status seekers. Non–status-seeking animals don't attract suitable mating partners, and they eventually exit the gene pool. Thus goods conveying high status remain extremely important, yet are out of reach for most of us. Nothing technology brings about will change that. Yes, one day we might reengineer our cognition to reduce or eliminate status competition. But until then, most people will have to live with the frustrations of technology's broken promise: That is, goods and services will be available to everybody at virtually no cost, but at the same time status-conveying goods will inch even further out of reach. That's a paradox of material progress.

Yes, luxury used to define things that made life easier: clean water, central heating, fridges, cars, TVs, smartphones. Today luxury tends to make your life harder. Displaying and safeguarding a Rauschenberg, learning to play polo and maintaining an adequate stable of horses, or obtaining access to the Pope are arduous undertakings. That doesn't matter; their very unattainability, the fact that these things are almost impossible for most people, is what matters.

As global wealth increases, nonreproducible goods will appreciate exponentially. Too much status-seeking wealth and talent is eyeing too few status-delivering goods. The price of nonreproducible goods is even more dependent on the inequality of wealth than on the absolute level of wealth in a society—further contributing to this squeeze.

The promise of technological progress can by definition not be kept. I think we should worry about the consequences, including a conceivable backlash to the current economic ecosystem of technology, capitalism, and free trade.

CLOSE OBSERVATION AND DESCRIPTION

URSULA MARTIN
Professor of computer science, Queen Mary, University of London

Take a potato from a bag of potatoes. Look at it closely. Yes, that's right, you really have to do this exercise. Now put it back with the other potatoes, mix them up, and see if you can find it again. Easy? Now try this with oranges. Still easy? Now recruit a friend, and describe, draw, or even photograph the potato in enough detail for them to pick it out. Or the orange.

There is a weed with a pink flower growing in my flower border, and something very similar, with a yellow flower, growing in the wall. These are fumitory (*Fumaria officinalis*) and corydalis (*Pseudofumaria lutea*), wildflowers of the poppy family, and yes, the field guide tells me that the former is commonly found in cultivated ground and the latter in limestone walls. To identify exactly which of the twelve or so U.K. fumitories I am holding in my hand requires me to consider the precise descriptive language of botany and count and measure inflorescences, peduncules, and sepals. The field guide helpfully provides hand-drawn colored illustrations, black-and-white diagrams of flower parts, and a backwards look-up key ("If the inflorescence is shorter than the peduncle and . . . then it is a . . .") to supplement the textual descriptions.

Once upon a time, such observation, description, and illustration were the bread and butter of professional and amateur scientists. My eight-volume flora, on heavy paper with lovely illustrations that are now collector's items, was well thumbed by the original owner, a 19th-century lady of leisure. It claims to be written for the "unscientific," but the content differs from a modern flora only by the inclusion of quantities of folklore, anecdotes, and literary references.

Darwin's books and letters are full of careful descriptions. The amateur struggling with a field guide may take comfort in reading how he frets over the difference between a stem with two leaves and a leaf with two leaflets. Darwin seems to have had a soft spot for fumitories, giving wonderfully detailed descriptions of the different varieties, whether and under what conditions they attracted insects and how the geometry and flexibility of the different parts of the flower affected how pollen was carried off by visiting bees. He was looking for mechanisms that ensured evolutionary variability by making it likely that bees would occasionally transfer pollen from one flower to another, giving rise to occasional crosses—analysis later reflected in the *Origin of Species*.

Shakespeare and his audiences not only knew their weeds, they knew their habitats, too, and the difference between arable land and permanent pasture. In *Henry V*, we hear how wartime neglect has changed the countryside so that arable land left fallow has been taken over by darnel, hemlock, and fumitory, and in the unmown meadows, "hateful docks, rough thistles, kecksies, burs" have replaced cowslip, burnet, and clover. Gardeners everywhere will sympathize.

Google can give a happy hour or so tracking down references to fumitory: Online data sources allow the capture and analysis of images, sightings, and geographic and other data in a way never before possible, informing the creation and testing of broader scientific hypotheses. But the analysis is only as good as the input provided—citizen science projects founder if the citizens are unable to do more than record a "pink flower" and a blurred mobile-phone image. No amount of image analysis or data mining can yet take the place of the attention and precision practiced by Darwin and thousands of other professional and amateur naturalists and ecologists.

So let's hear it for observation and description. Fumitory and corydalis may be for the advanced class. Start with the potato.

IMPACT

BRUCE HOOD
Director, Bristol Cognitive Development Centre, University of Bristol; author, The Self Illusion: How the Social Brain Creates Identity

As someone fairly committed to the death of our solar system and ultimately the entropy of the universe, I think the question of what we should worry about is irrelevant in the end. In any case, natural selection eventually corrects for perturbations that threaten the stability of environments. Nature will find a way, and ultimately all things will cease to be. So I could be glib and simply say, "Don't worry, be happy." Of course we are not wired that way, and being happy requires not worrying. My concern then, rather than worry, is how we go about science, and in particular the obsession with *impact*.

Up until the last century, science was largely the prerogative of the independently wealthy, who had the resources and time to pursue their passions for discovery. Later, large commercial companies would invest in research and development to gain the edge over competitors by innovating. The introduction of government funding for science in the early part of the 20th century was spurred by wars, economic depression, and disease. This not only broadened the scope of research by enabling much larger projects that were not motivated simply by profit but also created a new professional: the government-funded scientist.

In the U.K., the end of the 20th century was the golden period for funding. Since then, there has been significant shrinkage, in the West at least, of support for research in science. Today it is much harder to attract funding for research, as governments grapple with the world recession—unless, of course, that research generates economic wealth.

It used to be the case that for a research-grant application the results of any output were expected to be disseminated in publications or presentations at conferences—expectations that could be covered in a sentence or two. In the U.K. today (and I imagine this also true in the U.S.), a significant part of the application must address something called "pathways to impact." What does that mean, exactly?

According to the U.K. research council's own guidelines, it has to be a

> demonstrable contribution that excellent research makes to society and the economy. Impact embraces all the extremely diverse ways in which research-related knowledge and skills benefit individuals, organisations and nations by: fostering global economic performance, and specifically the economic competitiveness of the U.K.; increasing the effectiveness of public services and policy; enhancing quality of life, health and creative output.

This is not simply a box-ticking exercise. As part of the nationwide assessment of U.K. research known as the Research Excellence Framework (REF), impact features prominently in the equation. "What's the problem?" you might ask. Taxpayers' money funds research, and they need a return on their investment.

The first major problem is that it shifts the agenda away from scientific discovery to the application of science. I have witnessed in my own department in the past ten years that those who work on theoretical science are not as successful at procuring funding as those who work on application. Moreover, that application is primarily motivated by economic goals. Universities are being encouraged to form partnerships with industry to make up for the reduction in government funding. This is problematic for two reasons: The practices and agendas of industry conflict with those of the independent researcher; moreover, many important innovations were not conceived as applications and would probably not have emerged in an environment that emphasized commercial value. I would submit that focusing on impact is a case of putting the cart before the horse—or at least of not recognizing the value of theoretical work. We would be wise to remember Francis Bacon's advice that serendipity is a natural consequence of the pursuit of science.

Many of us work in areas that are difficult to fit into the impact framework. My own research is theoretical. When I'm asked to provide a pathway-to-impact statement, I rely on my experience of, and enjoyment in, delivering public lectures, because frankly the things that interest me do not obviously translate into impact that will foster economic performance. However, public engagement can be problematic, especially when one is addressing issues of concern. Most members of the general public—and, more important, the media that inform them—are not familiar with either the scientific method or statistics. This is one reason why the public is so suspicious of scientists, or finds them frustrating because they never seem to give a straight answer on such pertinent issues as vaccination or health risks. Most nonscientists do not understand explanations couched in terms of probability or complex, multifactorial interactions. Weekly headlines like "X Causes Cancer" or "The Discovery of Genes for X" reflect this need to simplify scientific findings.

Finally, most academics themselves have succumbed to the allure of impact. Every science journal has an impact factor, which is a measure of how often articles are cited. It is a reasonable metric, but it creates a bias in the scientific process by prioritizing those studies that are the most extraordinary. As we have witnessed in the past few years, this has led to the downfall of several high-profile scientists, who lost their jobs because they fabricated studies that ended up in high-impact journals. Why did they do this? Simply because you need impact in order to succeed. My concern is that impact is incompatible with good science because it distorts the process by looking for the immediate payoff and to hell with caution.

Maybe my concern is unwarranted. Science is self-correcting, and when the world comes out of recession we should see a return to the balance between theory and application. But then perhaps I should have been more alarmist; that way I'd probably have made more impact.

THE COMPLEX, CONSEQUENTIAL, NOT-SO-EASY DECISIONS ABOUT OUR WATER RESOURCES

GIULIO BOCCALETTI

Physicist, atmospheric & oceanic scientist; managing director, The Nature Conservancy

We should be worried about the state of water resources. I doubt there is a single current-affairs publication that has not addressed the "global water crisis." On balance, they raise legitimate concerns.

In the next twenty years, we will need to supply roughly 40 percent more water than we do today to support greater economic activity, from food to energy production. Because almost half of global food production comes from 20 percent of cultivated land under irrigation, it is unlikely that we are going to meet the food requirements of a growing and wealthier population without capturing, storing, and delivering more water.

The traditional solution to this is to build new infrastructure—reservoirs, dams, canals. This may very well be the right answer in some places, particularly in those countries that have yet to build much of their water infrastructure, but overall it may prove too expensive to be the only solution.

We must change the way in which we use water, doing more with less. Unfortunately, we do not have a great track record in increasing resource productivity. In 1967, the National Academy of Sciences established the Committee on Resources and Man to answer that neo-Malthusian question: Are we able to increase our resource productivity so as not to exhaust the planet? Or is our exponential growth a precursor of ecosystem collapse on a global scale? Almost fifty years later, the good news is that we are still here. The bad news is that, as envisioned by the committee, our increased productivity has not proved sufficient: In water, as in other resources, so far it has been outstripped by demand.

But concerns for water don't stop at issues of quantity. In developed countries, thousands of soluble chemical compounds are making their way into water bodies in trace concentrations. Pharmaceuticals (from anti-inflammatories to antidepressants), personal-care products, detergents, pesticides, various hydrocarbons—the list is long and growing. Whether or not some of these will turn out to have significant epidemiological consequences remains to be seen, but in most cases standard treatment technologies are not designed to intercept them.

We are not doing so well on the ecosystems front, either. I sometimes have the impression that freshwater conservation is perceived as a concern only by middle-aged accountants with a passion for fishing. It should not be. More than a quarter of all vertebrate species on the planet live in rivers. We have lost, proportionally, more species in freshwater ecosystems than in any other. That deterioration is also bad news for us. Biologically well-functioning rivers and adjacent ecosystems may be a necessary counterweight to increasing our use of water resources for economic activities. If we are to grow intensive irrigated agriculture and increase our fertilizer use, functioning riparian ecosystems are essential to mitigate their impact and ensure that costs for potable treatment don't grow unchecked. If we want management of flood risks to be more resilient in the face of climate change, we will have to increasingly blend hard infrastructure with functioning wetlands.

So we have cause for concern, no question about it. That said, when it comes to water, "worry" (particularly that of a catastrophic nature) is a singularly unhelpful sentiment. Water systems are not simply a feature of our landscape that needs to be protected from human activities. They are a foundational element of our societies and a basic infrastructure for our economies. That they are subject to so many competing and at times incompatible interests is an inevitable consequence of their critical importance in virtually everything we do.

Reams have been devoted to rousing alarms about the global water crisis, but the barrier to effective action is not one of conviction but of complexity. It is frightening that—outside of a restricted circle of practitioners—we have been unable to develop a fact-based, practical way of debating water issues in public. We don't need heated public debates on the likelihood of "water wars," or on the popular (mistaken) idea that water is an exhaustible resource. But we do need debates on the practical, consequential choices we have.

How should we think about the tradeoffs between food production and water scarcity? What options do we have on hydropower, and what are their costs? What does an economy consistent with the available resources look like? Precisely because managing water for multiple objectives is complex and solutions are always contextual, we must make sure that people know what questions to ask and understand the tradeoffs they will have to live with.

For many countries, ensuring enough water to sustainably satisfy the needs of a growing population and economy will be a complicated and expensive balancing act. Globally, $600 billion is spent each year on managing water, a sum comparable to what is spent on producing natural gas. And this does not include expenditure in other sectors—from agriculture to manufacturing—that influences the intensity of water use.

Historically, governments have financed most of those expenditures and negotiated the tradeoffs between competing objectives, as part of their ordinary administrative process. But today few public institutions have access to the necessary funds to pay for such investments, and many struggle to access affordable finance. Even when the money is there, the question of managing water resources is no longer just administrative. As we stretch what we can do with our limited resources, choices across the economy become increasingly interconnected: Industrial policy, energy, and agricultural choices become water choices, and the environmental outcomes we seek for our rivers have implications for jobs and economic development.

These are profoundly political and value-laden issues that deserve informed public debate. People must be able to debate about the tradeoffs they're asked to live with, just as they do—or should do—with other collective strategic issues, from energy to health care. If they don't, our economies, societies, and environment will inevitably be the poorer for it.

CHILDREN OF NEWTON AND MODERNITY

STUART A. KAUFFMAN

Founding director, Institute for Biocomplexity & Informatics, University of Calgary; author, Reinventing the Sacred: A New View of Science, Reason, and Religion

The great early sociologist Max Weber wrote, "With Newton we became disenchanted and entered Modernity." I believe that Weber was right: We remain disenchanted and are inarticulately lost in Modernity. Many of us seem to sense the end of something—perhaps a futile meaninglessness in our Modernity.

Was Weber right? And how is it based on science? Yes, Weber was right, and our disenchantment remains because, at least in part, we remain Newton's children.

Before Newton, in the previous two centuries in Europe, we had the black and white magi. Kepler was the last of the white magi, hoping the Platonic solids would state the orbits of the planets—only to surpass that, and Aristotle's certainty that orbits were perfect (thus circles), when he discovered that the planetary orbits were ellipses.

But the black magi sought occult knowledge, to stand Nature on her head and wrest their due, a misguided use of God's promise to Adam in Genesis. In those centuries, we lived with magic and were enchanted.

Newton changed everything with his three laws of motion, universal gravitation, and his invention of differential and integral calculus. Consider seven billiard balls rolling on a billiard table. What will they do? Newton taught us to measure initial conditions of positions and momenta of the balls, the boundary conditions of the table, then write the forces between the balls (and walls) using differential-equation forms of his laws of motion. Then, said Newton, we were to *integrate* his equations of motion to obtain the forever trajectories of the balls—ignoring friction for the moment. But integration is *deduction* of the consequences of the differential equations, and deduction is logical *entailment*. So Newton, in what I'll call the Newtonian Paradigm, gave us both classical physics and an entirely entailed becoming of physical systems.

With the Marquis de Laplace, a bit more than a century later, we obtain the view that were all the positions and momenta of all the particles in the universe known, a giant computer in the sky, the Laplacian Demon, could, using Newton's laws, calculate the entire entailed future and past of the universe. This is the birth of modern reductionism in physics, Weinberg's Dream of a Final Theory that will entail all that becomes in the universe.

Quantum mechanics does not change this fundamental view. In place of deterministic, entailed trajectories, we have, from the Schrödinger equation, the entailed trajectory of a probability distribution, obtained as the squares of the amplitudes of the Schrödinger linear wave equation. In modern physics—general relativity and quantum mechanics—all that becomes in the universe is entailed. Nothing novel can arise. And, profoundly, we remain disenchanted. There can be no natural magic.

From Newton we get the Enlightenment, our Age of Reason, thence the Industrial Revolution, and then the rise of Modernity. We remain disenchanted.

Beyond the Newtonian Paradigm

I claim that—at least for the living, evolving world; the evolving biosphere, human economy, legal systems, culture, and history—no laws at all entail the becoming of these worlds into their forever newly emerging but un-pre-statable "adjacent possible opportunities," which, in evolution, are not achieved by the "action" of natural selection. Nor in human life are these adjacent possible opportunities typically "achieved" by human intent.

Because these evolutionary processes typically cannot be pre-stated, the very phase space of biological, economic, cultural, and legal evolution keeps changing in un-pre-statable ways. In physics, we can always pre-state the phase space, hence can write laws of motion, hence can integrate them to obtain the entailed becoming of the physical system. But because in the evolution of life, and human life, the very phase space changes in un-pre-statable ways, we can write no laws of motion. Nor can we non-circularly pre-state the boundary conditions on this evolution, so we have neither laws of motion nor their boundary conditions and so cannot integrate the entailed trajectories of the laws of motion (which we do not have anyway).

To show why we cannot pre-state the evolution of the biosphere, I start in a strange place: Please list for me *all* the uses of a screwdriver. Well, to screw in a screw, open a can of paint, wedge a door open (or shut), stab an assailant, tie to a stick to make a fish spear, rent the spear to locals for 5 percent of the catch. . . .

Here seem to be the new and essential issues: (1) The uses of a screw driver are indefinite in number; (2) these uses, unlike the integers, are in no way naturally orderable.

But these two premises imply *that no effective procedure, or algorithm*, can list *all* the uses of a screw driver. This is the famous *frame problem* in algorithmic computer science, unsolved since Turing and his machine.

But all that has to happen in the evolution of a bacterium in, say, some new environment is that a molecular screwdriver "finds a use" that enhances the fitness of the bacterium and that there be heritable variance for that "use." Then natural selection will "pull out" this new use by selecting at the level of the bacterium, not the molecular screwdriver.

The profound implication of the newly selected bacterium with the molecular screwdriver is that this evolutionary step changes the very phase space of evolution in an *un-pre-statable way*. Hence we can write no laws of motion for this evolution, nor can we pre-state the niche boundary conditions noncircularly, so we could not even integrate the laws of motion we cannot write in the first place. Since we cannot list all the uses of the molecular screwdriver, we do not know the sample space of evolution.

Evolution of the biosphere and, *a fortiori*, of the human economy, legal systems, culture, and history, are *entailed by no laws at all*. True novelty can arise, beyond the Newtonian Paradigm broken beyond the watershed of life.

Re-enchantment, a path beyond Modernity, is open to us.

WHERE DID YOU GET THAT FACT?

VICTORIA STODDEN
Computational legal scholar; assistant professor of statistics, Columbia University

We are being inundated every day with computational findings, conclusions, and statistics. In op-eds, policy debates, and public discussions, numbers are presented with the finality of a slammed door. In fact we need to know how these findings were reached, so we can evaluate their relevance, their credibility, resolve conflicts when they differ, and make better decisions. Even figuring out where a number came from is a challenge, let alone trying to understand how it was determined.

This is important because of how we reason. In the thousands of decisions we make each day, seldom do we engage in a deliberately rational process anything like gathering relevant information, distilling it into useful knowledge, and comparing options. In most situations, standing around weighing pros against cons is a pretty good way to ensure rape, pillage, and defeat, either metaphorical or real, and miss out on pleasures in life. So of course we don't very often do it; instead, we make quick decisions based on instinct, intuition, heuristics, and shortcuts honed over millions of years.

Computers, however, are very good at components of the decision-making process that we're not: They can store vast amounts of data accurately, organize and filter it, carry out blindingly fast computations, and beautifully display the results. Computers can't (yet?) direct problem solving or contextualize findings, but for certain important sets of questions they are invaluable in enabling us to make much more informed decisions.

They operate at scales our brains can't, and they make it possible to tackle problems at ever greater levels of complexity.

The goal of better decision making is behind the current hype surrounding big data, the emergence of "evidence-based" everything—policy, medicine, practice, management, and issues such as climate change, fiscal predictions, health assessment, even what information you are exposed to online. The field of statistics has been addressing the reliability of results derived from data for a long time, with many successful contributions (for example, confidence intervals, quantifying the distribution of model errors, and the concept of robustness).

The scientific method suggests skepticism when interpreting conclusions and a responsibility to communicate scientific findings transparently so others may evaluate and understand the result. We need to bring these notions into our everyday expectations when presented with new computational results. We should be able to dig in and find out where the statistics came from, how they were computed, and why we should believe them. Those concepts receive almost no consideration when findings are publicly communicated.

I'm not saying we should independently verify every fact that enters our daily life—there just isn't enough time, even if we wanted to—but the ability should exist where possible, especially for knowledge generated with the help of computers. Even if no one actually tries to follow the chain of reasoning and calculations, more care will be taken when generating the findings when the potential for inspection exists. If only a small number of people look into the reasoning behind results, they might find issues, provide needed context, or be able to confirm their acceptance of the finding as is. In most cases, the technology exists to make this possible.

Here's an example. When news articles started appearing on the World Wide Web in the 1990s, I remember eagerly anticipating hot-linked stats—being able to click on any number in the text to see where it came from. More than a decade later, this still isn't routine, and facts are asserted without the possibility of verification. For any conclusions that enter the public sphere, it should be expected that all the steps that generated the knowledge are disclosed, including making the data they're based on available for inspection whenever possible and making available the computer programs that carried out the data analysis—open data, open source, scientific reproducibility.

Without the ability to question findings, we risk fooling ourselves into thinking we are capitalizing on the Information Age when we're really just making decisions based on evidence that no one, except perhaps the people who generated it, can actually understand. That's the door closing.

IS IDIOCRACY LOOMING?

DOUGLAS T. KENRICK

Professor of psychology, Arizona State University; author, Sex, Murder, and the Meaning of Life

The 2006 movie *Idiocracy* was hardly Academy Award material, but it began with an interesting premise: Given that there is no strong selection for high IQ in the modern world, people who are less intelligent are having more children than the more intelligent people. Extrapolating that trend for 500 years, the movie's producers depicted a world populated by numbskulls. Is this a possibility?

There are several causes for concern. To begin with, it is a correct assumption that natural selection is largely agnostic with regard to intelligence. We large-brained hominids like to think that all the information-crunching power in our hypertrophied cortexes will eventually allow us to solve the big problems of modern times, so that our descendants persist into the distant future. But it ain't necessarily so. Dinosaurs were a lot smarter than cockroaches, and Australopithecines were Einsteinian by comparison, yet the roaches have had a much longer run and are widely expected to outlast *Homo sapiens.*

Consider a few later phenomena:

1. Even correcting for other factors, people living in larger families have lower IQs.
2. In the modern world, less-educated people reproduce earlier and have larger families than highly educated people.
3. Less-educated people are more likely to hold conservative religious beliefs than are better-educated people.

4. Conservative religiosity is associated with opposition to birth control and abortion. The psychologist Jason Weeden has data suggesting that this is, in fact, close to the heart of the split between the liberal left and the conservative right.
5. Some conservative religions, such as the Church of Jesus Christ of Latter-day Saints, actively encourage large families.
6. Other conservative religions, such as the Roman Catholic Church, indirectly encourage large families by forbidding most means of family planning.
7. Larger families are likely to be poorer, and poverty triggers earlier puberty and earlier reproduction (a good deal of recent research suggests that this phenomenon is linked to biological life-history patterns and unfolds independent of any normative inputs from religions or local culture). These factors combine to ensure that poorer, less-educated young people are more likely to stay that way and produce the next (slightly larger) generation of poor, less-educated young people.
8. Well-educated intellectual types have smaller families these days, and because highly educated women wait longer to begin reproduction, they often miss their fertile window and have no children.

During the 20th century, IQ tended to generally increase (a phenomenon dubbed the Flynn Effect, after the researcher who discovered it). Various hypotheses have been advanced for this phenomenon, including better education and smaller families. But the factors I listed above could set a course in the opposite direction, flip-flopping the Flynn effect.

And there is another potential ironic twist. If the population of less-educated, religiously conservative individuals increases and continues to vote as they have been voting, funding for education and scientific research is also likely to decrease. A less-educated population could contribute not only to an upward shift in population size but also to a downward economic spiral, for reasons linked to some fascinating findings by Heiner Rindermann and James Thompson. These researchers examined the economic consequences of variations in IQ across ninety countries, analyzing the average IQ of each country's population as a whole, as well as the average IQ of the "intellectual elite" (the top 5 percent of the population), and the lowest 5 percent of the population. Just as countries differ in their distribution of wealth, they also differ in their distribution of IQ. Canada and the United States, for example, are identical in the average IQ of their smartest people (120), but the lowest 5 percent of Canadians are 5 points smarter than the lowest 5 percent of Americans (80 vs. 75).[e]

Rindermann and Thompson's analyses led them to this conclusion: Having a high-IQ intellectual class—lots of people with accomplishments in science, math, technology, and engineering—translates directly into more wealth for a country. To put it in purely economic terms, an increase of one IQ point among average citizens increases a country's average GDP by $229, whereas an increase of one IQ point in the intellectual elite is worth $468.

A high-performing intellectual class is also associated with better-developed and freer economic and political institutions, which in turn encourages more development of the country's "cognitive resources," in what Rindermann and Thompson call a "virtuous cycle." The free climate becomes a catalyst for creative productivity among the high-IQ innovators, who are thus liberated to rethink older ways of doing things and expand into new scientific arenas. This in turn inspires new technologies as well as newer and more efficient ways of doing business and a better climate in which to grow the next generation of innovators.

So to the extent that a growing anti-intellectual portion of the population manages to cut funds for education and for scientific research, they effectively sabotage the system that feeds what has been the world's most productive "human capital" machine. According to my colleagues from other countries, the American educational system has an unimpressive reputation up through the university level but is regarded as the top of the heap when it comes to training at the highest level—with people around the world desperate to come to the U.S. to get the best PhD training on the planet. Thus, slashing funding for higher education and scientific research (much of which is conducted by our best PhD students at major universities) seems a policy destined to undermine the country's economic health in long-lasting ways.

Policies that undercut the intellectual upper crust therefore undermine economic growth and indirectly contribute to the economic threat that inspires poorer, less-educated people to reproduce earlier and more prolifically.

THE DISCONNECT BETWEEN NEWS AND UNDERSTANDING

GAVIN SCHMIDT
Climatologist, NASA's Goddard Institute for Space Studies

We are surrounded by complexity. Issues demanding our attention—health-care reform, climate change, the Arab Spring—have a historical context, multiple viewpoints, clashes of diverse values, and a bewildering set of players with their own agendas. The news provides a source for what's happening right now: voting results, who made what speech, how many people died, etc. While imperfect, the news industry mostly delivers on its duty to provide information on what's "new," and with the advent of social media platforms and aggregators like Google News, it has never been easier to stay up-to-date.

But much of what we need to understand a situation is not "new." We need a deeper knowledge of the context to inform our understanding of why the new events have occurred. The situation in Afghanistan makes no sense without an appreciation of the culture and history of the region. The latest warning of a future climate effect makes no sense unless you understand how we know anything about how the climate operates and how it has already changed. Understanding the forces driving the Arab Spring requires a background in the breakup of the Ottoman Empire and the responses to the Colonial adventurism that followed. Unfortunately, this context is not in the least bit newsworthy.

The gap between new and old is widening, and that should be profoundly worrying. It's as if we had a populace that was well informed about the score of a game but knew nothing about the rules and, worse, had no inclination to find credible sources to explain them. Public discussions often devolve to mere tribalism; it is far easier to base decisions on who supports what than to delve into an issue yourself. Any efforts to make it easier to access depth and context must therefore be applauded and extended. New online tools can be developed to scaffold information by providing entry points appropriate for any level of knowledge. Context buttons alongside online searches could direct the interested to the background information. But unless we start collectively worrying about this, nothing will change, and our society's ability to deal with complexity in a rational way will continue to decline.

SUPER-AIs WON'T RULE THE WORLD (UNLESS THEY GET CULTURE FIRST)

ANDY CLARK

Philosopher, chair in logic & metaphysics, University of Edinburgh; author, Supersizing the Mind

The last decades have seen fantastic advances in machine learning and robotics. These are now coupled with the availability of huge and varied databases, staggering memory capacities, and ever faster and funkier processors. But despite all that, we should not fear that our Artificial Intelligences will soon match and then rapidly outpace human understanding, turning us into their slaves, toys, pets, or puppets.

For we humans benefit from one gigantic and currently human-specific advantage: the huge yet nearly invisible mass of gradually accrued cultural practices and innovations that tweak and pummel the inputs that human brains receive. Those gradually accrued practices are, crucially, delicately keyed to the many initial biases—including especially biases for sociality, play, and exploration—installed by the much slower processes of biological evolution. In this way, a slowly accumulated mass of well-matched cultural practices and innovations ratchets up human understanding.

By building and then immersing ourselves in a succession of designer environments, such as the human-built worlds of education, structured play, art, and science, we restructure and rebuild our own minds. These designer environments are purpose-built for creatures like us, and they "know" us as well as we know them. As a species, we refine them again and again, generation by generation. It is this iterative restructuring and not sheer processing power, memory, mobility, or even the learning algorithms themselves that is the final (but crucial) ingredient in the mental mixture.

To round it all off, if recent arguments by Oxford psychologist Cecilia Heyes are correct, many of our capacities for cultural learning are themselves cultural innovations, acquired by social interactions rather than flowing directly from biological adaptations. In other words, culture itself may be responsible for many of the mechanisms that give the cultural snowball the means and momentum to deliver minds like ours.

Why does this mean that we should not fear the emergence of superintelligent AI anytime soon? The reason is that only a well-structured route through the huge mass of available data will enable even the best learning algorithm (embodied perhaps in multiple, active, information-seeking agents) to acquire anything resembling a real understanding of the world—the kind of understanding needed even to generate the goal of dominating humankind. Such a route would need to be specifically tailored to the initial biases, drives, and action capacities of the machines themselves. If the slow coevolution of body, brain, biases, and an ever-changing cascade of well-matched cultural practices is indeed the key to advanced cognitive success, we need not fear the march of the machines. For the moment, there is simply nothing in the world of the AIs that looks set to provide that kind of enabling ladder.

"Deep Learning" algorithms are now showing us how to use artificial neural networks in ways that come closer than ever before to delivering learning on a grand scale. But we probably need "deep culture" as well as deep learning if we are ever to press genuine hyperintelligence from the large databases that drive our best probabilistic learning machines.

That means staged sequences of cultural practices, delicately keyed to the machines' own capacities to act and communicate and tuned to the initial biases and eco-niche characteristic of the machines themselves. Such tricks ratchet up human understanding in ways that artificial systems have yet to even begin to emulate.

POSTHUMAN GEOGRAPHY

DAVID DALRYMPLE
Grantee, Thiel Foundation, Nemaload; Program in Biophysics, Harvard University; research affiliate, Synthetic Neurobiology group, MIT Media Lab

When the value of human labor is decimated by advances in robotics and artificial intelligence, serious restructuring will be needed in our economic, legal, political, social, and cultural institutions. Such changes are being planned for by approximately nobody. This is rather worrisome.

If every conceivable human job can be done better by a special-purpose machine, it won't make any sense for people to have jobs with corporations to earn wages they exchange for goods and services. This isn't the first time an entire paradigm of civilization has become obsolete; the corporation itself is only 500 years old. Before that, a "job" meant a single project; craftsmen and merchants traveled from city to city, as businesses unto themselves.

The corporation is useful because it can bring hundreds, thousands, or even millions of people together, working toward a common purpose with central planning. Countries are useful for similar reasons but at a larger scale and with the added complexity associated with enforcement (of laws, borders, and "the national interest"). But as technology advances, the context that gives power to those sorts of institutions will shift dramatically. As communication becomes cheaper, higher-speed, and more transparent, central planning becomes a less appropriate tool for bringing people together. One of the most centrally-planned organizations in history, the Soviet Union, was brought down by the fax machine, which enabled citizens to bypass the state-run media. The corporation system is much more adaptable, but we're already seeing it struggle with the likes of "hacktivist" group Anonymous.

More important, future technologies will be used to bring people together in new, fluid structures with unprecedented productivity. New sorts of entities are emerging that are neither countries nor corporations, and in the not-too-distant future such agents will become less "fringe," ultimately dominating both industry and geopolitics.

Individual humans will eventually fade even from the social world. When you can literally wire your brain to others, who's to say where you stop and they begin? When you can transfer your mind to artificial embodiments and copy it as a digital file, which one is you? We'll need a new language, a new conceptual vocabulary of everything from democracy to property to consciousness, to make sense of such a world.

One possible way to begin developing such ideas is reflected in my title. It's a bit of wordplay, referring to three disparate intellectual movements: transhumanism, which predicts the emergence of technologically enhanced "posthumans"; human geography, which studies humans' relations with one another and the spaces they inhabit; and posthumanism, a school of criticism revisiting the usual assumption that individuals exist. These fields have much to learn from each other, and combining them would be a good start to addressing these issues.

For example, transhumanist stories about the future (like most stories) feature individuals as characters. They may be able to communicate ideas to one another "telepathically," in addition to other new capabilities, like being able to move one's mind from one embodiment to another (thus avoiding most causes of death). But they are still recognizable as people. However, with high-bandwidth brain interfaces, as posthumanist critics tell us, that might not be the case.

Human geography tends to ignore technological trends unless they're incremental—or at least easily described in terms of existing concepts. Transhumanist technologies would dramatically change the geographic picture (as a simple example, solar-powered humans wouldn't need agriculture). These geographic changes, in turn, would have significant consequences for the trajectory of transhumanist technology and society.

Posthumanism is highly abstract. Perhaps predictably, thinkers interested in questions like "Do individuals exist?" or "What is the meaning of identity?" tend to be uninterested in questions like "In light of this, how will equity markets need to evolve?" They also tend to be generally suspicious of technology, often citing the distorting influence of the mass media. If they're aware of transhumanism at all, they criticize it for its individualism and then move on, rather than looking deeper and seeing the possibility of totally new types of identity structures with these future capabilities. Finally, although posthumanists do talk about economies and societies, it's usually without reference to real-life data or even historical examples.

Personally, I have little doubt that the next paradigm of civilization will be a change for the better. I'm not worried about that. But the transition from here to there might be painful if we don't develop some idea of what we're getting into and how it might be managed.

BEING TOLD THAT OUR DESTINY IS AMONG THE STARS

ED REGIS
Science writer; coauthor (with George Church), Regenesis: How Synthetic Biology Will Reinvent Nature and Ourselves

In 2012, NASA and DARPA jointly funded a 100-Year Starship program, the goal of which was to achieve human interstellar flight within the next 100 years. On September 13, 2012, the 100-Year Starship (100YSS) project held the first of its planned annual public symposiums, in Houston, Texas. Here, about 100 scientists, social scientists, educators, journalists, and miscellaneous others gathered to witness a series of scientific presentations outlining schemes by which human beings could, just possibly, leave planet Earth behind, travel to the stars, and establish a new "Earth 2.0" in another solar system, all within a century.

Traveling to the stars, say many of its advocates, is our preordained destiny as a species. As proponent Cameron Smith puts it in "Starship Humanity" (*Scientific American*, January 2013): "the concept of a Space Ark, a giant craft carrying thousands of space colonists on a one-way, multigenerational voyage far from Earth" is "technologically inevitable."

Far from being technologically "inevitable," the fact is that such a voyage is not even known to be technologically *possible*. For one thing, the distances to even the closest extrasolar stars are unimaginably vast. The nearest star, Proxima Centauri, is 4.22 light-years (24,800,000,000,000 miles) from Earth. Even if we were to travel as fast as the *Voyager 1* spacecraft, which is now receding from us at 38,120 mph, it would take an interstellar craft more than 73,000 years to reach that destination.

But traveling at significantly faster speeds requires prohibitive amounts of energy. If the starship were propelled by conventional chemical fuels at even 10 percent of the speed of light, it would need for the voyage a quantity of propellant equivalent in mass to the planet Jupiter. To overcome this limitation, champions of interstellar travel have proposed "exotic" propulsion systems, such as antimatter, pi-meson, and space-warp propulsion devices. Each of these schemes faces substantial difficulties of its own: For example, since matter and antimatter annihilate each other, an antimatter propulsion system must solve the problem of confining the antimatter and directing the antimatter nozzle in the required direction. Both pi-meson and space-warp propulsion systems are so very exotic that neither is known to be scientifically feasible.

Indeed, these and other such schemes are really just mathematical abstractions, not working systems: They are major extrapolations from states of matter that exist today only at nano levels. (Making even tiny amounts of antimatter, for instance, requires huge accelerators, at stupendous cost.) Still other systems depend on wild possibilities such as making use of extra dimensions that are not known to exist and physical forces or influences that are not known to be real or are sheer flights of the imagination (such as altering the value of Hubble's constant to make the universe smaller).

Even if, by some miracle, suitable propulsion systems became available, a starship traveling at relativistic speeds would have to be equipped with sophisticated collision-detection-and-avoidance systems, given that a high-speed collision with something as small as a grain of salt would be like encountering an

H-bomb. Star voyagers face further existential threats in the form of prolonged exposure to ionizing radiation, boredom, alienation from the natural environment, the possible occurrence of a mass epidemic, the rise of a charismatic leader who might derail the whole project, crew mutiny, religious factionalism, and so on. It is far more likely, therefore, that an interstellar voyage will mean not the survival but the death of its crew.

Apart from all of these difficulties, the more important point is that there is no good reason to make the trip in the first place. If we need an Earth 2.0, then the moon, Mars, Europa, or other intra-solar-system bodies are far more likely candidates for human colonization than are planets light-years away.

So, however romantic and dreamy it might sound, and however much it might appeal to one's youthful hankerings of "going into space," interstellar flight remains a science-fictional concept—and with any luck it always will be.

COMMUNITIES OF FATE

MARGARET LEVI
Political scientist, Bacharach Professor of International Studies, University of Washington; author, Consent, Dissent, and Patriotism

We all live in communities of fate; our fates are entwined with others in ways we perceive and ways we cannot. Our individual and group actions often have consequences, sometimes unforeseen, that affect others in significant ways. Moreover, there are those, beyond our families with whom we feel entangled, whose interests and welfare we perceive as tied to our own. In the lingo of the Industrial Workers of the World (the Wobblies), "an injury to one is an injury to all." Who is included in the "one" and who in the "all" differs among us, and therein lies the rub.

How we understand our community of fate matters for the perils the world faces today and into the future. That so many people choose to live in ways that narrow the community of fate to a limited set of others and define the rest as threatening to their way of life and values is deeply worrying, because this contemporary form of tribalism, and the ideologies that support it, enables them to deny complex and cross-cutting interdependencies—local, national, and international—and ignore their own role in creating threats to their own and others' well-being. Climate change and religious zealotry are among the well-documented examples of individual and group actions that have significant spillover effects. Perhaps most alarming is our collective failure to incorporate future generations into our communities of fate. The implications for almost all public policies are enormous.

But there is hope. It is possible to expand the community of fate well beyond the network of those we personally know. We continually make decisions that have consequences for others; there is now evidence that if we understood those consequences and expanded our community of fate, we might make different choices. For example, we buy cheap goods without acknowledging that they may entail labor that is badly paid or treated. As research by Michael Hiscox and Jens Hainmueller details, more information and a different framing can transform the consumption patterns of at least some buyers.

Jaron Lanier, in *Edge*, has argued that the "new moral question" is whether our decisions are self-interested or take others into account. Whom an individual includes in her community of fate (what Lanier labels "circle of empathy") is a moral question, yes, but it is also an empirical question. Organizational and institutional factors can explain variation and transformation in the community of fate. They can enable individuals to revise their understanding of the consequences of their actions and change the boundaries of their communities of fate. Of critical importance are the group's governance arrangements and the procedures for enforcing them. All groups have rules and norms defining membership and appropriate behavior; membership organizations, including government, codify those rules and norms into formal institutions. How the community of fate—and not just membership—is defined requires leaders committed to principles they are willing to uphold in unforeseen circumstances, punishment of violations (to make these commitments credible), and provision of services, security, and other benefits members expect.

But the community can be narrow or broad. To expand the community of fate beyond the actual members of an organization depends on the desire and ability of leaders to convince members that their welfare is tied up with a larger set of others, often unknown others. Success further requires providing new information, in a context of strong or potential challenge to its veracity. Organizations most effective in expanding the community of fate are those allowing members to question what they're being told and seek additional information if need be. In practice, this implies some form of participatory governance or bottom-up democracy.

The broadening of the community with whom individuals feel mutual dependence is generally the result of factors promoting collective action: incentives, punishments, monitoring over time. But with a push from the behavioral economists, we now recognize the equal importance of appeals to fairness and the creation of emotional attachments. At least some people, some of the time, will make individual sacrifices in order to combat unfairness. What Elisabeth Wood labels "the pleasure of agency" can motivate actions that serve a larger whole. As Ernst Fehr, Samuel Bowles, and others find, implicit and emotional incentives can lead to complementarities rather than substitution of intrinsic and extrinsic motivations.

So the good news is that communities of fate can be expanded to include strangers and future generations. But there is a problem here—and two additional sources of worry. First, we are far from figuring out how to use our scientific knowledge about our interdependencies to make more people aware of their shared welfare. How do we extend what we have learned from the lab and case studies to a larger set of institutions and organizations? Second, and even more troubling, the mechanisms that draw people out of their narrow self-serving behavior can be used for good or ill. Terrorists and other zealots of the religious, nationalist, or political kind believe they are serving a larger collective good, at least for those within their definition of the collective to be served, and many sacrifice for their cause. How do we inhibit these groups and promote the governments and organizations that serve a wider public? That is, indeed, a problem worth worrying about.

WORKING WITH OTHERS

STEPHEN M. KOSSLYN

Psychologist; director, Center for Advanced Study in the Behavioral Sciences, Stanford University; coauthor (with William Thompson & Giorgio Ganis), The Case for Mental Imagery

ROBIN S. ROSENBERG

Clinical psychologist; author, Superhero Origins: What Makes Superheroes Tick and Why We Care

As the 21st century proceeds, tasks facing our species will become increasingly complex. Many problems that in an earlier era might have been easily addressed by one person will now require a sophisticated set of abilities contributed by different people. The individual contributions must be complementary; the whole must be more than the sum of its parts.

This much seems obvious. But no part of contemporary formal education—at any point from kindergarten through postgraduate work—is designed to teach people how to interact effectively with other people in goal-oriented groups. When such a group functions well, it synergizes the talents and abilities of its members. But at present such synergy occurs because of a lucky combination of people who happen to have complementary skills and abilities relevant to the task at hand, and who happen to be able to interact effectively. It's not obvious how best to compose a group to facilitate such synergy. But most people don't seem aware that there's a problem here.

For example: Many people who interview job applicants think they're good at picking the "right" applicant—that they know how to pick appropriate employees based not just the content of an applicant's answers but also on his or her nonverbal behavior. But it has been repeatedly shown that interviewers who rely on intuition and "feeling" generally are *not* good at picking job applicants.

So, too, with selecting people to work together in goal-oriented groups. People have intuitions about how to assign individuals to groups and how to organize them, but decisions based on such intuitions are not necessarily any better than chance. Relying on the luck of the draw won't be very effective as task-oriented groups face increasingly complex challenges. We must overcome such intuitions. We need to realize that understanding how best to select the right people for the right group is a hard problem—and so is understanding how they should interact most effectively.

To compose a group that can effectively tackle a complex problem, we need to know (1) how to analyze the nature of tasks, in order to identify the necessary skills and abilities; (2) how to identify such skills and abilities in individuals; and (3) how different sorts of people can interact most effectively when working on a particular sort of task. Much research will be required to crack these problems (and such research is already under way), but the results will not be widely applied as long as people don't recognize the nature of the problems and why they are important.

Science can do better than intuition—but we first must understand that intuition isn't good enough. And this isn't intuitively obvious.

GLOBAL COOPERATION IS FAILING AND WE DON'T KNOW WHY

DANIEL HAUN
Leader, Comparative Cognitive Anthropology Group, Max Planck Institute for Evolutionary Anthropology

More than ever, we depend on the successful cooperation of nations in making planetary-scale decisions. Many of today's most dire problems cannot be solved by actions taken within single countries but will find solutions only if the global community joins forces. However, global cooperation is mostly failing, even if the problem is well defined and the options for action fairly well understood. Why? One answer is that humans' narrow pursuit of their own interests will trump any possibility for collective action. This explanation is as insufficient as it is pessimistic. If this were the heart of the problem, failure would be inevitable. We are not going to change human nature in time.

• The good news is that inherent human self-interest, while not exactly making things easy, is not the only reason that global cooperation is failing. As well as being inherently self-interested, humans, it turns out, are inherently cooperative. From an early age, children are able and, even more important, highly motivated to interact with others in order to accomplish common goals. Kids enjoy cooperating. Moreover, other primates seem to have many of the basic abilities necessary for cooperation. Hence, cooperation appears to have been part of humans' biological history for a long time, with many of the necessary abilities and motivations heritable. Even human adults choose cooperation over defection, unless they are given too much time to think about it.

When making decisions quickly, people's gut reaction is to cooperate. We cooperate much more than we might predict based on self-interest alone. Cooperation is as much part of human nature as is self-interest, so it's not a foregone conclusion that global cooperation will always fail. The challenge is to build a context for global cooperation that brings out our cooperative side.

• The bad news is that we have no idea how to do that. That's because we don't know how cooperation works when you go from small groups of people to a planet-wide level. We come to this problem at crunch time with empty hands. We should worry about that.

• What we do know is how to build a context that promotes cooperation among limited groups of individuals (local cooperation). For example, local cooperation works well when the chance for reciprocation is high and the cooperators trust one another. Individuals cooperate better when the channels for communication are effective, the cooperators think of themselves as similar to one another, and the system allows for cheaters to be punished. What we don't know is how these factors operate in a global-cooperation scenario. But isn't global cooperation just like local cooperation except with lots more individuals? Can't we just take what we know about local cooperation and extrapolate to a group of 7 billion? The reason global cooperation fails could simply be that reciprocation, trust, communication, punishment, and interindividual similarity are all compromised on that scale. If this were the heart of the problem, again, failure would be inevitable. Luckily it isn't. Because that's not how global cooperation works.

• Yes, no one can hide from the consequences of failing global cooperation. Yes, everybody will in some way become part of global solutions. But what is the actual scale at which global cooperation happens? Global cooperation involves multiple layers of decision-making instances. Decisions are frequently delegated to representatives of larger communities to coordinate action at local, regional, national, and global levels. Thus, while the consequences of what we call "global cooperation" are in fact global, the decision-making process itself is done in relatively small groups of people. So why doesn't it work? What are the differences between local cooperation, as we know and experience it every day, and global cooperation, if the size of the group of cooperators is roughly the same?

- The difference is that the individuals who are interacting in a global-cooperation scenario respond to a different, more complex problem space. International representatives who carry out the bulk of negotiations interact in principle like individuals, but they carry the burden of speaking for a much larger constituency. These people act both as members of a small group of experts and as representatives of the interests of thousands or millions of people. The interests of the masses are one of the factors taken into account by those making decisions on their behalf. This added layer of responsibility might fundamentally change the structure of cooperation. We desperately need to know how.
- Besides their own and their constituents' stakes in mitigating global climate change, these representatives also have to worry much more about the popularity of their decisions than is typical in everyday local cooperation. While we all worry to some extent about the social consequences of our decisions, the careers of the representatives in global cooperative scenarios depend on their popularity—the reelection problem. The consideration of reelection seriously narrows any representative's decision space in an already increasingly difficult scenario. Hence the institutions

we built to succeed in solving local cooperative dilemmas are potentially inadequate for global cooperation.

• While we understand local cooperation quite well, these factors and likely many others turn global cooperation into another beast altogether. While we battle to strap a saddle on its back, the chances of success are minimal unless we learn how humans interact in these new and extraordinary scenarios and use that information to build structures for global cooperation that bring out the cooperative side of human nature.

THE BEHAVIOR OF NORMAL PEOPLE

KARL SABBAGH
Writer, television producer; author, Remembering Our Childhood: How Memory Betrays Us

The most worrying aspect of our society is the low index of suspicion about the behavior of normal people. In spite of the doctrine of original sin that permeates Christianity, the assumption most of us hold about most people in everyday life is that they are not, on the whole, criminals, cheats, mean-spirited, selfish, or on the lookout for a fast buck. Bad behavior is seen as something to be noticed, reported on, and analyzed, whereas people who do not lie and cheat are taken for granted. Good behavior is seen as the default mode for humans, and bad behavior is seen as "aberrant," even though, from self-knowledge as well as experiments like Stanley Milgram's, we know that "normal" people are not always saints.

This unwillingness to believe the worst of people permeates society and harms us in all sorts of ways. The most egregious current example, of course, concerns bankers and financiers, who have shown that only the most severe constraints on their activities would stop them from filching our purses and grabbing huge salaries or severance payments that are usually rewards for failure. And it is precisely those constraints that the institutions resist most strongly, promising after each one of their crimes is exposed that self-regulation will prevent the next. But any daily newspaper will show countless examples of individuals who demonstrate that when they (we?) can get away with something, they will.

There is much psychological research into the nature of evil. This usually proceeds from the assumption that people are naturally good and tries to explain why some depart from this "norm." Isn't it time we took the opposite view and looked into why some people, perhaps not many, are "good"? If you look hard, you can find examples of these. Whistleblowers, for example, who cannot stand by while bankers fiddle, doctors cover up mistakes, priests abuse children, or statesmen cheat on their expenses. Paradoxically, instead of their good deeds being welcomed, whistleblowers are often ostracized, even by people whose behavior is not—at least overtly—meretricious but who feel that reporting illicit behavior is itself distasteful.

Just as there is emerging evidence of a biological basis for political beliefs, left-wing vs. right-wing, perhaps we should be looking for a biological basis of goodness. So many of the world's problems have at their root the propensity of humans—and, indeed, nations—to behave in ways that try to maximize benefits to themselves at the expense of others. Even attempts to instigate solutions to climate change are bedeviled by personal and corporate selfishness.

What we should be worried about, therefore, is that science is missing out on a possible alternative solution to many of our problems. It should be considering how to make more people "good" rather than trying to understand what makes people "bad."

METAWORRY

BRIAN KNUTSON

Associate professor of psychology & neuroscience, Stanford University

I worry about worry. Specifically, is our worry aimed at the right targets? The adaptive value of worry is that it helps us avoid death without having to experience it first. But worry can save us only when directed at actual threats and only by eliciting action (avoidance). Although worry seems to be caused by external factors, it isn't. The neural worry engine is always on, looking for its next target, like Freud's free-floating anxiety.

The ancestral environment probably tuned our worry engines, since individual differences in levels of worry show significant heritability (around 50 percent) and are normally distributed (most people experience moderate rather than minimal or excessive worry). This bell-shaped distribution implies that over generations those who worried too little died (or were eaten), whereas those who worried too much failed to live (or reproduce). Thus our forebears' menu of environmental threats likely selected an optimal level of worry. Less appreciated is the notion that the ancestral environment selected not only the level but also the targets of worry. Consider common targets of phobias, such as public judgment, snakes, spiders, heights. Unless you live in the jungles of New Guinea, these are probably not the existential threats worthy of your worry engine. Leading causes of death in the United States typically are more "boring" (heart disease, cancer, stroke, accidents), encouraged by more proximal causes (smoking, alcohol consumption, poor diet, lack of exercise, cars, firearms). This "worry gap" between imagined and actual threats suggests that our worry is often directed at the wrong targets.

My *metaworry*, or worry about worry, is that actual threats are changing much more rapidly than they have in the ancestral past, which could widen the worry gap. Humans have created much of this environment with our mechanisms, computers, and algorithms that induce rapid, disruptive, and even global change. Financial and environmental examples spring to mind—witness crashes of global finance bubbles and the rise in global temperature over the past decade. Not only are these changes rapid with respect to an evolutionary time frame but they plausibly result from human causes. I worry that our worry engines will not retune their direction to focus on these rapidly changing threats fast enough to take preventive action.

We could try closing the widening worry gap with data. If we could calculate the relative risk of death by snakes, spiders, cars, and guns, we could compare the discrepancy between what we worry about (e.g., spiders) and what statistically causes more deaths (e.g., cars). Then we could try to swap our lower-ranked worry with a higher-ranked worry. We could even sketch the outlines of software that facilitates a better reallocation of worry. (I'm looking at you, mobile-application developers.) If we're going to worry about something, it might as well actually threaten our existence.

Another option for closing the worry gap involves policy change. Governments undoubtedly collect valuable data on the relative potency of various threats so as to direct their limited resources toward reducing the most pressing dangers. But information alone is not enough; it must motivate behavioral change. Thus, laws (with enforceable sanctions) are sometimes needed to transform this information into action. Unfortunately, in the case of global threats, governments must coordinate their laws. This is not impossible; it has happened in the past, when governments came together to ban chlorofluorocarbons in order to stop atmospheric-ozone depletion. It may happen in the future with respect to carbon sequestration and global climate change. But it requires a massively coordinated and continued effort.

I'm advocating *metaworry* rather than *hyperworry*. Hyperworry feeds on itself: Escalating worrying about worrying could fuel a positive-feedback loop, ending in a fearful freeze. Since the brain has limited energy, we should probably view worry as a resource to be conserved and efficiently allocated. Beyond increasing the worry level, metaworry implies redirecting worry based on information. Retuning our ancient worry engines may be difficult, but not impossible. Consider laws requiring the use of seatbelts while driving. What I find miraculous is not that these laws have become mandatory in most U.S. states, nor even that they reduce fatalities as predicted, but rather that the laws now grab me at a visceral level. Twenty-five years ago, I would never have thought twice about driving without a seatbelt. Now, when I can't buckle my seatbelt I feel uneasy and tense. In a word, I worry—and seek to close the gap.

MORBID ANXIETY

JOEL GOLD

Psychiatrist; clinical associate professor of psychiatry, NYU Langone Medical Center

Worry in and of itself can be extremely corrosive to our lives. The things we worry about tend to be distinct from those problems that grab our attention, stimulate us, and mobilize us to action. It is useful to recognize a potential or imminent danger and then do something about it; it is quite another matter to ruminate in a state of paralysis. Taking scientific, political, or personal action to address climate change is progressive. Lying sleepless in bed thinking about the melting polar caps does nothing but take a mental and physical toll on the insomniac.

The very word "worry" connotes passivity and helplessness: We worry about some event in the future and hope it works out. And the question we are posed this year implies that much of how we worry is irrational. I'm confident that many of the responses addressing what we *shouldn't* be worrying about will reassure readers because they will be specific and supported by the evidence. "Don't worry, be happy" bromides are of no use; notice that people who are told to "relax" rarely do.

Moreover, there is a large segment of the population for whom the serenity prayer really doesn't cut it. They can't distinguish between the obstacles in life they might actively address, thereby diminishing worry, and those for which there is no salutary behavior and are therefore best put out of mind. These people worry not only about problems that have no solutions but also about circumstances that to others may not seem problems at all. For this group, irrational worry becomes pathological anxiety. I am writing about the tens of millions of Americans who suffer from anxiety disorders.

I currently treat people in my psychiatric practice who are often derisively called "the worried well." They are most certainly worried, but they are not at all well. No one would question the morbidity associated with the irrational fears of the psychotic patients I once treated at Bellevue Hospital: the fear that they were being poisoned; that thoughts were being inserted into their minds; that they had damned the Earth to Hell. Yet trying to rationally explain away someone's phobia—aside from the use of structured cognitive behavioral therapy—is no more useful than trying to talk someone out of his delusion.

People with anxiety disorders suffer terribly: The young man with OCD, tormented by disturbing imagery and forced to wash his hands for hours until they bleed; the veteran haunted by flashbacks of combat, literally reexperiencing horrors physically survived; the professional who must drive hundreds of miles for fear of flying; the former star athlete, now trapped, unable to leave her home, fearing she might have another panic attack at the mall or while driving over a bridge. And the numbers are staggering: Anxiety disorders are the most commonly diagnosed psychiatric disorders, afflicting nearly 20 percent of all adult Americans each year and almost 30 percent over the course of their lives. Those suffering from phobias, panic, generalized anxiety disorder, PTSD, and the rest are more likely to be depressed, more prone to substance abuse, and more likely to die younger. And before death arrives, whether sooner or later, enjoyment of life is vitiated by their anxiety. Pleasure and anxiety do not co-occur.

We all want to avoid the biological and social risk factors for anxiety, but for tens of millions of Americans and hundreds of millions worldwide, it is too late. Once morbid anxiety takes hold, only psychotherapy and/or psychopharmacotherapy will suffice. Worry descends upon the worrier like a fever. Without appropriate treatment, that febrile anxiety burns away at the soul. With such treatment, the fever may break. Only then can the worried become well.

THE LOSS OF OUR COLLECTIVE COGNITION AND AWARENESS

DOUGLAS RUSHKOFF

Media theorist; documentarian; author, Present Shock: When Everything Happens Now

We should be worried about the decline of the human nervous system. We should be worried that something—likely environmental but possibly more subtle than that—is hampering our ability to parent new human beings with coherently functioning perceptual apparatuses. We should worry about what this means not just for our society's economic future but for the future of our collective cognition and awareness as a species.

Consider the rising proportion of our youth who are classified as "special needs." By current count, one in eighty-eight children has an autism spectrum disorder—and one in fifty-four boys. Eight percent of children between three and seventeen years old have a learning disability. Seven percent have ADHD. And these figures are rising steadily. The vast majority of children with spectrum disorders also suffer from "co-morbidities," such as anxiety disorder, sensory-processing disorder, intellectual disabilities, and social disorders. These special-needs children will likely need special education and then special care and special caretakers for the rest of their lives. Someone has to pay for this.

Moreover, as the numbers of children affected by these impairments increase, the amount of resources remaining to raise and educate the "neurotypical" members of our communities go down. The more children requiring $100,000 and up of special-needs education each year, the less money remains and the larger the regular classes must become. There are of course many social remedies for these problems, from increased community support and involvement to radical rethinking of education itself. But the less well educated we are, the less prepared we are tackle these issues.

And spectrum disorders are just the most visible evidence of neural breakdown. We must also consider the extensive use of SSRIs and other mood-enhancement drugs. Unlike psychedelics and other psychosocial learning substances, SSRIs were designed for continuous use. By treating stress, anxiety, and even transient depression as chronic conditions, the doctors prescribing these medications (influenced by the companies peddling them) are changing people's neurochemistry in order to dampen their responses to real life. The fact that 23 percent of American women between the ages of forty and fifty-nine take SSRIs may have less to say about this group's propensity for depression than it does about the way our society currently responds to women between forty and fifty-nine. To drug the victim may be merciful on some level but only paralyzes our collective self-regulation as a culture and a species.

So on the one hand, we should be worried about a future in which there are only a few of us left to keep the lights on, caring for a huge population of neurally-challenged adults. We are seeing the beginnings of this in a media-influenced society of imbecilic beliefs, road rage, and inappropriate reactions to stimuli. The reasonable people, on whom we depend for restaurants, hospitals, and democracy itself to function, seem to be dwindling in numbers, replaced by those with short tempers, inferiority complexes, and an inability to read basic social cues. On the other hand, if all this bothers us, we're supposed to take medication to alter our perceptions of and responses to social phenomena that should rightly make any healthy person depressed. And the number of people who have chosen to pharmaceutically limit their emotional range should be a concern—because they are no longer bearing witness to our collective reality. There are fewer of us left alarmed enough to take the action necessary.

Not to mention that all these drugs end up in the water supply and beyond, likely leading to increases in spectrum disorders among children. (Taking SSRIs during pregnancy, for instance, leads to double the probability of delivering a child on the autism spectrum.) So we come full circle, increasingly incapable of doing anything about this feedback loop, or even caring about it. We're soaking in it.

But there's an even greater concern. We should worry less about our species losing its biosphere than losing its soul. Our collective perceptions and cognition are our greatest evolutionary achievement. This is the activity that gives biology its meaning. Our human neural network is in the process of deteriorating and our perceptions are becoming skewed, both involuntarily and by our own hand. And all that most of us in the greater scientific community can do is hope that somehow technology picks up the slack, providing more accurate sensors, faster networks, and a new virtual home for complexity.

We should worry that such networks won't be able to function without us; we should also worry that they will.

WORRYING ABOUT CHILDREN

ALISON GOPNIK

Professor of psychology, UC Berkeley; author, The Philosophical Baby: What Children's Minds Tell Us About Truth, Love, and the Meaning of Life

Thinking about children, as I do for a living, and worrying go hand in hand. There is nothing in human life so important and urgent as raising the next generation, and yet it feels as if we have very little control over the outcome. British prime minister Stanley Baldwin once accused the press of having "power without responsibility—the prerogative of the harlot throughout the ages." Perhaps it's appropriate that the prerogative of the mother is just the opposite of the harlots: We moms have responsibility without power, a recipe for worry if ever there was one. But as a scientist as well as a mother, I worry that much of our current worry about children is misdirected. We worry a lot about the wrong things, and we don't worry nearly enough about the right ones.

Much modern middle-class worry stems from a fundamentally misguided picture of how children develop. It's the picture implicit in the peculiar but now ubiquitous concept of "parenting." As long as there have been *Homo sapiens*, there have been parents. Human mothers and fathers, and others as well, have taken special care of children. But the word "parenting" first emerged in America in the 20th century and became common only in the 1970s. This particular word comes with a picture, a vision of how we should understand the relations between grown-ups and children. "To parent" is a goal-directed verb. It describes a job, a kind of work. The goal is to shape your child into a particular kind of adult—smarter or happier or more successful than others. And the idea is that some set of strategies or techniques will accomplish this. So contemporary parents worry endlessly about whether they are using the right techniques and spend millions of dollars on books or programs that are supposed to provide them.

This picture is empirically misguided. "Parenting" worries focus on relatively small variations in what parents and children do—co-sleeping or crying it out, playing with one kind of toy rather than another, more homework or less. There is very little evidence that any of this makes much difference to the way children turn out in the long run. Nor does there seem to be any magic formula for making one well-loved and financially supported child any smarter or happier or more successful as an adult than another.

The picture is even more profoundly misguided from an evolutionary perspective. Childhood is one of the most distinctive evolutionary features of human beings; we have a much longer childhood than any other primate. This extended childhood seems, at least in part, to be an adaptation to the variability and unpredictability of human environments. The period of protected immaturity we call childhood gives humans a chance to learn, explore, and innovate without having to plan, act, and take care of themselves at the same time. And empirically we've discovered that even the youngest children have truly extraordinary abilities to learn and imagine, quite independent of any conscious parental shaping. Our long protected childhood arguably enables our distinctive human cognitive achievements.

The evolutionary emergence of our extended childhood went hand in hand with changes in the depth and breadth of human care for children. Humans developed a triple threat when it comes to care. Unlike our closest primate relatives, human fathers began to invest substantially in their children's care; women lived on past menopause to take care of their grandchildren; and unrelated adults—"alloparents"—kicked in care, too. In turn, children could learn a variety of skills, attitudes, knowledge, and cultural traditions from all those caregivers. This seems to have given human children a varied and multifaceted cognitive toolkit that they could revise and refine to face the various unpredictable challenges of the next generation.

So the evolutionary picture is that a community of caregivers provides children with two essential ingredients allowing them to thrive. First, adults provide an unconditionally nurturing and stable context, a guarantee that children will be safe and cared for. That secure base frees them to venture out to play, explore, and learn—to shape their own futures. Second, adults provide children with a wide range of models of acting in the world, even contradictory models of acting. Children can exploit this repertoire to create effective ways of acting in unpredictable and variable environments—and eventually to create new environments. This is very different from the "parenting" picture, where particular parental actions are supposed to shape children's adult characteristics.

This brings me to the stuff we don't worry about enough. While upper-middle-class parents worry about whether to put their children in forward- or backward-facing strollers, more than one in five children in the United States are growing up below the poverty line, and nearly half the children in America grow up in low-income households. Children, and especially young children, are more likely to live in poverty than any other age group. This number has increased substantially during the past decade. More significantly, these children face not only poverty but also a more crippling isolation and instability. It's not just that many children grow up without fathers: They grow up without grandparents or alloparents either, and with parents who are forced to spend long hours at unreliable and low-paying jobs. Institutions haven't stepped in to fill the gap. We still provide almost no public support for child care. We pay parents nothing and child-care workers next to nothing.

Of course, we've felt for a long time the moral intuition that neglecting children is wrong. But more recently, research into epigenetics has helped demonstrate just how the mechanisms of care and neglect work. Research in sociology and economics has shown empirically how significant the consequences of early experience can be. The small variations in middle-class parenting make little difference. But providing high-quality early childhood care to children who would otherwise not receive it makes an enormous and continuing difference up through adulthood. The evidence suggests that this isn't just a matter of teaching children particular skills or kinds of knowledge—a sort of broader institutional version of parenting. Instead, children who have a stable, nurturing, varied early environment thrive in a wide range of ways: better health, less propensity to crime, more successful marriages. That's just what we'd expect from the evolutionary story. I worry more and more about what will happen to the generations of children who don't have the uniquely human gift of a long, protected, stable childhood.

THE DEATH OF MATHEMATICS

KEITH DEVLIN

Executive director, H-STAR Institute, Stanford University; author, The Man of Numbers: Fibonacci's Arithmetic Revolution

Are we about to see advances in mathematics come to an end? Until last year, I would have said no. Now I'm not so sure. Given the degree to which the advances in science, engineering, technology, and medicine that created our modern world have all depended on advances in mathematics, if such advances were to end then it's hard to see anything ahead for society other than stagnation, and perhaps decline. How likely is this? I don't know. The escape clause in this year's *Edge* Question is that word "should." Had the question been "What *are* we worried about?" I suspect that any honest answer I could give would have focused on personal issues of health, aging, and mortality. Mathematics—dead or alive—would not have gotten a look.

Sure, there are many things going on in society that disturb me. But it's a consequence of growing older to view changes in society as being for the worse. We have been conditioned by the circumstances prevailing in our formative years, and when those circumstances change—as they must and should, since societies are living entities—we find it disconcerting. Things inevitably seem worse than they were in our childhood, but that reflects the fact that as children we simply accepted things as they were. To let a disturbed sensation as an adult turn into worry would be to adopt an egocentric view of the world. That said, we are reflective, rational beings with a degree of control over our actions both individual and collective, and it's prudent for us to read the signs and judge if a change in direction is called for.

One sign I came across unexpectedly last year was a hint that mathematics as we know it may die within a generation. The sign was, as you may suspect, technological, but not the kind you probably imagine. Though a stream of commentators in the 1960s and 1970s lamented that the rise of calculators and computers was leading to a generation of mathematical illiterates, all that was happening was a shift in focus in the mathematics that was important. In particular, a high arithmetical ability, crucial to successful living for centuries, suddenly went away—just as basic crop-growing and animal-rearing skills went away with the onset of the Industrial Revolution. In place of arithmetic skills, a greater need for algebraic thinking arose. (That in many cases the need was ill met by being taught as if it were arithmetic does not eliminate the new importance of algebraic skills in today's world.) There's no denying that technological advances change the kind of mathematics that is done. When calculating devices are available to everyone on the planet (North Korea excepted), which will almost certainly be the case in the next ten years or so, the world is unlikely to ever again see the kinds of discoveries made in earlier eras by mathematical giants such as Fermat, Gauss, Riemann, and Ramanujan. The notebooks they left behind show they spent many hours carrying out calculations in longhand as they investigated primality and other properties of numbers. This led them to develop such a deep understanding of numbers that they could formulate profound conjectures, some of which they or others were subsequently able to prove.

There are likely to be properties of numbers that will never be suspected, once everyone has access to powerful computing technology. (On the other hand, there's a gain, since the computer has spawned what is called experimental mathematics, where massive numerical simulations give rise to another kind of conjecture—ones that likely would not have been discovered without powerful computers. I described this phenomenon in *The Computer As Crucible*, a book with Jonathan Borwein a few years ago.) So technology can certainly change the direction of mathematical discovery. But can it cause its death? In September last year, I caught a glimpse of how this could occur. I was giving one of those (instantly famous) Stanford MOOCs, teaching the basic principles of mathematical thinking to a class of 64,000 students around the world. Since the course was on university-level mathematical thinking, not computation, there was no possibility of machine-graded assignments. The focus was on solving novel problems for which you don't have a standard procedure available—in some cases, constructing rigorous proofs of results. This kind of work is creative and symbolic and can be done only by covering sheets of paper (sometimes several sheets) with symbols and diagrams, sometimes using notations you devise specially in order to solve the problem at hand.

In a regular university class, I or my teaching assistants would grade the students' work, but in a MOOC that's not feasible, so I made use of a method called calibrated peer evaluation, whereby the students graded one another's work. To facilitate the anonymous sharing, I asked students to take good-quality smartphone photos of their work, or scan their pages into PDF, and upload them to the course Web site, where the MOOC software would organize distribution and track the grades. Soon a few students posted questions on the MOOC discussion forum, asking if they could type up their work in LaTeX, a mathematical typesetting program widely used in advanced work in mathematics, physics, computer science, and engineering. I said they could. In fact, I had used LaTeX to create the MOOC resources on the Web site, to create the weekly assignment sheets, and to self-publish the course textbook.

Now, LaTeX is a large complex system with an extremely long and steep learning curve, so I was not prepared for what happened next. From that first forum post on, hardly anyone submitted their work as an image of a handwritten page! Almost the entire class (more precisely, the MOOC-typical 10 percent who were highly active throughout the entire course) either went to tortuous lengths to write mathematics using regular keyboard text or else mastered sufficient LaTeX skills to do their work that way. The forum thread on how to use LaTeX became one of the largest and most frequently used in the course. I shudder to think of the amount of time my students spent on typesetting—time that would have been far better spent on the mathematics itself.

We have, it seems, become so accustomed to working on a keyboard and generating nicely laid-out pages that we are rapidly losing (if, indeed, we have not already lost) the habit—and love—of scribbling with paper and pencil. Our presentation technologies encourage form over substance. But if (free-form) scribbling goes away, then I think mathematics goes with it. You simply cannot do original mathematics at a keyboard. The cognitive load is too great. The increasing availability of pens that record what is being written may, I suppose, save the day. But that's not substantially different from taking a photo or scan of a page, so I'm not so sure that's the answer. The issue seems to be the expectation that our work should meet a certain presentation standard. In a world dominated by cheap sophisticated presentation technologies, paper-and-pencil work may go the way of the dodo. And if that happens, mathematics will no longer advance. As a living, growing subject, it will die. RIP mathematics? Maybe. We will likely know for certain within twenty years.

SHOULD WE WORRY ABOUT BEING UNABLE TO UNDERSTAND EVERYTHING?

CLIFFORD PICKOVER

Author of the Pickover Trilogy: The Medical Book, The Physics Book, *and* The Math Book

I used to worry that our mathematical and physical descriptions of the universe grow forever but our brains and language skills remain entrenched. Some of our computer chips and software are becoming mind-numbingly complex. New kinds of mathematics and physics are being discovered or created all the time, but we need fresh ways to think and understand.

I used to worry that we will understand less and less about more and more. For example, in the last few years, mathematical proofs have been offered for famous problems in the history of mathematics, but the arguments have been far too long and complicated for experts to be certain they are correct. Mathematician Thomas Hales had to wait five years before expert reviewers of his geometry paper, submitted to the journal *Annals of Mathematics*, finally decided that they could find no errors and that the journal should publish Hales' proof, but only with the disclaimer saying they were not certain it was right! Moreover, mathematician Keith Devlin has admitted in the *New York Times* that "The story is that mathematics has reached a stage of such abstraction that many of its frontier problems cannot be understood even by the experts."[f]

Israeli mathematician Doron Zeilberger recently observed that contemporary mathematics meetings were venues where few people could understand one another. The "burned-out" mathematicians just ambled from talk to talk, where "they didn't understand a word." Zeilberger wrote in 2009, "I just came back from attending the 1052nd AMS [American Mathematical Society] (sectional) meeting at Penn State, last weekend, and realized that the Kingdom of Mathematics is dead. Instead, we have a disjoint union of narrow specialties. . . . Not only do [the mathematicians] know nothing besides their narrow expertise, they don't care!" Incidentally, Zeilberger considers himself an ultrafinitist, an adherent of the philosophy denying the existence of the infinite set of natural numbers (the ordinary whole numbers used in counting). More startling, he suggests that even very large numbers don't exist—say, numbers greater than 10 raised to the power of 10 raised to the power of 10 raised to the power of 10. In Zeilberger's universe, when we start counting—1, 2, 3, 4, etc.—we can seemingly count forever; but eventually we will reach the largest number, and when we add 1 to it, we return to zero!

At the risk of further digression into the edges of understanding, consider the work of Japanese mathematician Shinichi Mochizuki. Some of his key proofs are based on "inter-universal Teichmüller theory." As he develops future proofs based on this mathematical machinery, which was honed over decades in many hundreds of pages, how many humans could possibly understand this work? What does it even mean to "understand" in contexts such as these? As mathematics and subatomic physics progress in the 21st century, the meaning of "understanding" obviously must morph, like a caterpillar into a butterfly. The limited, wet human brain is the caterpillar. The butterflies are our brains aided by computer prosthetics.

Should we be so worried about being unable to understand subatomic physics, quantum theory, cosmology, or the deep recesses of mathematics and philosophy? Perhaps we can let our worries slightly recede and just accept our models of the universe when they're useful. Today we use computers to help us reason beyond the limitations of our intuition. Computer experiments are leading mathematicians to discoveries and insights never dreamed of before the ubiquity of computers. Computers and computer graphics allow mathematicians to discover results long before they can prove them formally, and they open entirely new fields of mathematics. The educator David Berlinski once wrote, "The computer has . . . changed the very nature of mathematical experience, suggesting for the first time that mathematics, like physics, may yet become an empirical discipline, a place where things are discovered because they are seen." W. Mark Richardson fully understands that scientists and mathematicians must learn to live with mystery. In a 1998 article in *Science*, he wrote:

> As the island of knowledge grows, the surface that makes contact with mystery expands. When major theories are overturned, what we thought was certain knowledge gives way, and knowledge touches upon mystery differently. This newly uncovered mystery may be humbling and unsettling, but it is the cost of truth. Creative scientists, philosophers, and poets thrive at this shoreline.

THE DEMISE OF THE SCHOLAR

DANIEL L. EVERETT
Linguistic researcher; dean of arts & sciences, Bentley University; author, Language: The Cultural Tool

On April 22, 1961, the futurist and inventor R. Buckminster Fuller offered a futuristic vision for higher education, a vision he later published as *Education Automation: Freeing the Scholar to Return to His Studies.* His idea was that the population explosion and the emerging revolution in information technology called for a massive scaling-up of higher education. Perhaps the best recommendation from his book is that universities should "[g]et the most comprehensive generalized computer setup with network connections to process the documentaries that your faculty and graduate-student teams will manufacture objectively from the subjective gleanings of your vast new world- and universe-ranging student probers." His view was that this would lead to improvements in society and universities across the board, leaving scholars largely free of teaching and able to dedicate their time almost exclusively to research.

Massive Online Open Courses, MOOCs, seem on the surface to have achieved Fuller's vision. Yet even as MOOCs reach more students with high-quality lectures than any format in history, they simultaneously threaten the survival of scholarship. The reason is that rather than engaging the population in research and development via mass access to teacher-scholars, MOOCs lead us closer to the vo-tech-ization of higher education. Instead of leading people to deep thought, new ideas, and useful applications, MOOCs too often facilitate the cheaper earning of credentials (which some online educators already refer to as "badges," reminiscent of the Boy Scouts) that certify skill acquisition. And along with this shift in the goals and methods of higher education, the disappearance of scholarship as a profession seems ever more probable. This worries me. It should worry you.

The unexpected transmogrification of Fuller's dream into a potential nightmare may eliminate scholarship and the livelihood of the scholar because the vast majority of scholarship arises from "teacher-scholars"—men and women who enlighten their research via their students and inform their teaching by their research. Though MOOCs are unlikely to threaten elite universities, they will lead to many closings of smaller liberal arts colleges and branch campuses of state universities. Bill Gates predicts that a college education, through MOOCs, will soon cost only $2,000 and that "place-based activity in that college thing will be five times less important than it is today."[g] And with this loss of importance of colleges will come the demise of the teacher-scholar—the promoter of knowledge and scientific thinking. MOOCs will replace many universities, the employers of scholars, by online certificates.

The teacher-scholar is a moderately paid individual who models problem solving and a cultivation of thirst for knowledge. The home in which this modeling takes place, the modern college or university, is the descendant of the academy of Plato. The greatest accomplishments of Western civilization have come from university and college students past and present, the disciples of professionals who have taken them under their wing, taught them in classes, worked with them in research labs, and talked about the glories of learning over a beer (or a coffee) in campus pubs and student unions.

A few miles from my home in the greater Boston area are some of the finest universities in the world. When I look at the schedules of departments in the dozens of universities in this region, I see listings for nightly lectures, brown-bag lunches, lab reports, book launches, and debates. And I know that these are only part of the peripatetic learning that takes place in America's brick-and-mortar institutions. Replacing these experiences is not currently within the abilities or the objectives of MOOCs.

The driving force behind universities as they have come to be defined in the United States and other countries has been the idea of present and future discovery—learning new insights from the past, new cures for human ailments, new methods for thinking about problems, new ways of understanding value and values, new flows of capital and labor, and new methods of quantifying and interpreting the knowledge so laboriously attained. Yet this vision of the university as an incubator of discovery is dying. As a generation, we are acquiescing to the misinterpretation of universities as certifiers of job readiness rather than as cultivators of curiosity and fomenters of intellectual disruption of the status quo. The students who aspire to "find themselves" and change the world are being replaced by the students who want to find a job. The tragedy is that people believe that these objectives represent a stark choice rather than a compatible conjunction. Universities prepare us for both. For now.

The post–Internet idea that all knowledge should be cheap and quick fuels the explosion of interest in MOOCs and the view of education as the transmission of skill sets. If learning for its own sake is a luxury we can no longer afford, then society will not allow professors to "return to their research" as Fuller urged. Why employ so many professors or maintain so many student unions and so many libraries after the automation of education?

Learning is hard slogging helped by cups of hot chocolate, mugs of brew, and conversations with people who have slogged or are slogging themselves. It is the individual's labor as a member of a larger community of learners. Humans have achieved culture—the transgenerational communication and development of learning. One of the greatest values of modern societies is the strange and beautiful notion that it is worth several years of the life of its young to participate in communities of scholars. Not merely the wealthy young but all the academically dedicated youth of an entire country.

Fuller's vision was to improve teaching via professional video lectures interspersed with illustrative material—videos within videos. We have achieved this. But his corollary ideas that this technology would allow scholars to dedicate more time to research while simultaneously involving a larger percentage of the world's population in the process of discovery are threatened by the very technology he foretold. If scholars make their living as teacher-scholars and technology eliminates the need for most scholars to teach, then there will be few sources of income for scholars. And if students are more interested in finding employment than finding truth, then I worry how the economy can fulfill its vital need for scholarship, as students request primarily MOOCs and other classes that immediately enhance employability.

Fuller's prescient idea of education automation failed to anticipate the economic and intellectual consequences of its own success—the demise of the very scholar he expected it to protect. That should worry us all.

SCIENCE IS IN DANGER OF BECOMING THE ENEMY OF HUMANKIND

COLIN TUDGE

Writer & biologist; author, The Link: Uncovering Our Earliest Ancestor

We—scientists and the world in general—should be worried because:

Science has become increasingly narrow-minded—materialistic, reductionist, and inveterately anthropocentric, still rooted philosophically in the 18th century.

Science is increasingly equated with high tech. "Vocation" used to mean a deep desire to engage and seek out the truth, insofar as human beings are able. Now it means getting a job with Monsanto.

Britain's Royal Society has become a showcase for high tech, and high tech is perceived as the means to generate wealth—and this is deemed to be "realistic," the way of the world, not only acceptable but virtuous.

Science, in short, is in danger of losing its integrity and its intellectual independence—of becoming the handmaiden of big business and the most powerful governments. Since we cannot assume that increasing wealth and top-down technical control are good for the human race (or our fellow creatures), science, for all its wonder and all its achievements, is in danger of becoming the enemy of humankind.

Worst of all, perhaps, is that those in high places—both in science and in politics—don't seem to realize that this is happening.

They do not listen to criticism. "Public debate" means a one-way flow of information *de haut en bas.* The status quo, for all its obvious flaws and, indeed, its horrors, is seen as "progress." But those who are doing well out of it, including those intellectuals who have chosen to go along with the act, cannot even begin to consider that they could be wrong.

All this, by all measures, is a tragedy.

ILLUSIONS OF UNDERSTANDING AND THE LOSS OF INTELLECTUAL HUMILITY

TANIA LOMBROZO
Assistant professor of psychology, UC Berkeley

My spelling has deteriorated as automated spell-checking has improved. With a smartphone at my fingertips, why bother doing multidigit multiplication in my head? And I can't say whether GPS and navigation software have made my mental maps any more or less accurate.

These are just some of the skills that have been partially outsourced from many minds, thanks to recent technological advances. And it won't be long before important aspects of our social and intellectual lives follow suit. As interfaces improve and barriers between mind and machine break down, information of all kinds isn't simply at our fingertips but seamlessly integrated with our actions in the world.

Forfeiting skills like spelling, navigation, and even certain kinds of social knowledge to our gadgets doesn't worry me much. What does worry me is the illusion of knowledge and understanding that can result from having information so readily and effortlessly available. Research in the cognitive sciences suggests that people rely on a variety of cues in assessing their own understanding. Many of these cues involve the way information is accessed. For example, if you can easily call an image, word, or statement to mind, you're more likely to think you've successfully learned it and to refrain from effortful cognitive processing.

Such fluency is sometimes a reliable guide to understanding, but it's also easy to fool. Just presenting a problem in a font that's harder to read can decrease fluency and trigger more effortful processing, with the surprising consequence that, for example, people do better at logical deduction and tricky word problems when the problems are presented in a less readable font. It follows that smarter and more efficient information retrieval via machines could foster dumber and less effective information-processing in human minds.

Consider another example: The educational psychologist Marcia Linn and her collaborators have studied the "deceptive clarity" that can result from complex scientific visualizations of the kind that technology in the classroom and online education are making ever more readily available.[h] Such clarity can be deceptive because the transparency and memorability of the visualization is mistaken for genuine understanding. It isn't just that when the visualization is gone the understanding goes with it, but that genuine understanding was never achieved in the first place.

People suffer from illusions of understanding even absent fancy technology, but current trends toward faster, easier, and more seamless information retrieval threaten to exacerbate rather than correct any misplaced confidence in what we truly comprehend. These trends also threaten to undermine some natural mechanisms for self-correction. For example, people are often more accurate in assessing their own understanding after explaining something to someone else (or even to themselves) or after a delay. Removing social interaction and time from informational transactions could therefore have costs when it comes to our ability to track our understanding.

Are technological advances and illusions of understanding inevitably intertwined? Fortunately not. If a change in font or delay in access can attenuate fluency, then a host of other minor tweaks to the way information is accessed and presented can surely do the same. In educational contexts, deceptive clarity can be partly overcome by introducing what psychologist Robert Bjork calls "desirable difficulties," such as varying the conditions under which information is presented, delaying feedback, or engaging learners in generation and critique—all of which disrupt a false sense of comprehension. And some of the social mechanisms that help calibrate understanding, such as explanation, can be facilitated by the marriage of information technologies and social media.

But avoiding illusions of understanding, and the intellectual hubris they portend, will require more than a change in technology—it will require a change in our expectations and behavior. We have to give up the idea that fast and easy access to information is always better access to information.

THE END OF HARDSHIP INOCULATION

ADAM ALTER

Psychologist; assistant professor of marketing, Stern School of Business, NYU; author, Drunk Tank Pink: And Other Unexpected Forces That Shape How We Think, Feel, and Behave

When psychologists ask people to tackle a new mental task in the laboratory, they begin with a round of practice trials. As the novelty of the experience wears off, participants develop a tentative mastery over the task, no longer wasting their limited cognitive resources on trying to remember which buttons to push, or repeatedly rehearsing the responses they're expected to perform. Just as vaccines inoculate people against disease, a small dose of practice trials prepares participants for the rigors of the experiment proper.

The same logic explains how children come to master the mental difficulties confronting them as they grow into adulthood. Trivial hardships inure them to greater future challenges that might otherwise defeat them but for the help of these experiential scaffolds. A child who remembers his mother's phone number is thereafter better equipped to memorize other numerical information. Another who routinely performs mental arithmetic in math class develops the skills needed to perform more complex mental algorithms. A third who sits bored at home on a rainy day is forced to devise new forms of entertainment, meanwhile learning the rudiments of critical thinking.

Unfortunately, these crucial experiences are declining with the rise of lifestyle technologies. The operation of iPhones and iPads is miraculously intuitive, but their user-friendliness means that children as young as three or four can learn to use them. Smartphones and tablets eradicate the need to remember phone numbers, perform mental calculations, and seek new forms of entertainment, so children of the 21st century never experience the minor hardships attending those tasks. They certainly derive other benefits from technology, but convenience and stimulation are double-edged swords, also heralding the decline of hardship inoculation. Today's children might thus be poorly prepared for the more difficult tasks that will meet them as time passes.

What's particularly worrying is not that today's children will grow up to be cognitively unprepared but the question of what the trend portends for their children, grandchildren, and so on. The "ideal" world—the one that looks more and more like the contemporary world with each passing generation—is the same world that fails to prepare us to memorize, compute, generate, elaborate, and, more generally, to think. We don't yet know which cognitive capacities will be usurped by machines and gadgets, but the range will widen over time, and the people who run governments, businesses, and scientific enterprises will be the poorer prepared because of this foregone vaccination.

INTERNET SILOS

LARRY SANGER
Cofounder of Wikipedia & founder of Citizendium

We should be worried about online silos. They make us stupid and hostile toward one another.

Internet silos are news, information, opinion, and discussion communities dominated by a single point of view. Examples are the *Huffington Post* on the left and *National Review Online* on the right, but these are only a couple of examples—and not the worst, either. In technology, *Slashdot* is a different kind of silo—of geek attitudes.

Information silos in general are nothing new and not limited to the Internet; talk radio works this way, churches and academia are often silos, and businesses and organizations study how to avoid a silo culture. But Internet communities are particularly subject to a silo mentality because they are virtually instant: They have no history of competing diverse traditions and they are self-selecting, thus self-reinforcing. The differences between online communities tend to be stark. That's why there are so many silos online.

It shouldn't be surprising that silos are fun and compelling for a lot of us. They make us feel like we belong. They reinforce our core assumptions and give us easily digestible talking points, obviating the need for difficult individual thought. They appeal to our epistemic vanity and our laziness.

That's one of the problems. Silos make us overconfident and uncritical. Silos worry me because critical knowledge—the only kind there is, about anything difficult—requires a robust marketplace of ideas. Silos give too much credence to objectively unsupportable views that stroke the egos of their members. In a broader marketplace, such ideas would be subjected to much-needed scrutiny. Silos are epistemically suspect. They make us stupider. They might be full of (biased) information, but they make us less critical-thinking and hence lower the quality of our belief systems.

It can be social suicide to criticize a silo from within the silo, while external criticism tends to bounce off, ignored. So silos become hostile to dissent, empowering fanatics and power-seekers at the expense of the more moderate and the truth-seekers. Silos also alienate us from one another—even from friends and family members who don't share our assumptions—because it is fun, and too easy, to demonize the opposition from within a silo. The rise of the Internet seems to correlate with the rise in the late 1990s and 2000s of a particularly bitter partisan hostility that has, if anything, gotten worse and made reaching political compromise increasingly unpopular and difficult. This threatens the health of the republic, considering that compromise has been the lifeblood of politics since the founding.

My solution? For one thing, you can do your part by regularly visiting the opposition and showing them in conversation how reasonable you can be. There's little more upsetting to a silo than infiltration by an intelligent, persistent individual.

THE NEW AGE OF ANXIETY

GARY KLEIN

Senior scientist, MacroCognition LLC; author, Sources of Power: How People Make Decisions

The *Edge* Question this year asks us to identify new worries, but I was not aware that we were running short of things to worry about. Just the reverse—we already have too many threats to keep us up at night. And that's what worries me. It seems we have entered a new Age of Anxiety. If the problem were internal, it could be treated with anti-anxiety drugs. Unfortunately, the problem is external, in the form of the ever expanding list of fears that science generates and the media are happy to amplify. Bad news sells. We listen more carefully to news reports about a possible blizzard than to a forecast of mild and sunny weather. And so the science/media complex is happy to feed our fears with all kinds of new threats.

I worry that the *number* of things we need to worry about keeps growing. The science/media complex is inventive at discovering all kinds of threats to our food and our water supplies—delighted to warn about the deterioration of the environment, declining fertility rates, new carcinogens, physical and mental health issues, and so on. The more novel the threat the better, because new dangers avoid our tendency to habituate to scare stories after they've been broadcast for a while. Very few old worries get retired. A few diseases, such as smallpox, may be conquered. But even there the science/media complex keeps us worried that terrorists might get hold of smallpox samples or re-create the disease in a lab and wreak havoc on a world that no longer gets smallpox immunization so our vulnerability to smallpox may be increasing, not shrinking.

I worry that the *shrillness* of worries keeps escalating. In a sea of worries, a new one stands out only if its consequences are apocalyptic. If it doesn't threaten our civilization, it won't get much airtime. The pressure is on scientists and media specialists to show that the new issue is not just dangerous but highly dangerous. It cannot merely be contagious; there has to be a means for it to mutate, or perhaps attach itself to a common vector, posing the threat of a deadly new plague. And shrillness isn't only about the consequences, it's also about the need to act immediately. To command our attention, the science/media complex has to show that this new problem should jump to the top of our priority list of worries. The threat has to be close to a tipping point beyond which it will become uncontrollable.

And I worry about the *proposed remedies* for each new danger. To be worth its salt, a new threat has to command rapid and extreme reactions. These reactions have to start immediately, eliminating our chance to evaluate them for unintended consequences. The more over-the-top our fears, the more disproportionate the reactions and the greater the chances of making things worse, not better.

I hesitate to raise the issue, because I'm just adding to the problem, but I do think it's something worth worrying about, and I don't see any easy way to counter this new Age of Anxiety. The science/media complex keeps ramping up, continually finding novel dangers, more threats to keep us up at night—and then we have to worry about the consequences of sleep deprivation. It never stops.

DOES THE HUMAN SPECIES HAVE THE WILL TO SURVIVE?

DAVE WINER

Blogging & RSS software pioneer; editor, Scripting News *Weblog*

Until a few generations ago, the human species was dealing with the following question: "Do we have what it takes to survive?" We answered that question with the invention of heat, plumbing, medicine, and agriculture. Now we have the means to survive, but do we have the will?

This is the 800-pound question in the middle of the room.

NEURAL DATA PRIVACY RIGHTS

MELANIE SWAN
Systems-level thinker; futurist; principal, MS Futures Group; founder, DIYgenomics

A worry not yet on the scientific or cultural agenda is neural-data privacy rights. Not even biometric-data privacy rights are in purview yet, which is surprising, given the personal data streams amassing from quantified self-tracking activities. There are several reasons why neural-data privacy rights could become an important concern: First, personalized neural data streams are already available from sleep-monitoring devices, and this could expand to eventually include data from eye-tracking glasses, continuously worn consumer EEGs, and portable MRIs. At some point, the validity and utility of neural data may be established with correlation to a variety of human health and physical and mental performance states. Despite the sensitivity of these data, security may be practically impossible. Malicious hacking of personal biometric data could occur and would need an in-kind response. There could be many required and optional uses of personal biometric and neurometric data for which we would need different permissioning models.

Personal biometric data streams are growing as people engage in quantified self-tracking with smartphone applications, biomonitoring gadgets, and other Internet-connected tools. The adoption of wearable electronics (smartwatches, disposable patches, augmented eyewear) could hasten this and might even outstrip tablets (now the most quickly adopted electronics platform). This could allow the unobtrusive collection of vast amounts of previously unavailable objective metric data—including not only biometrics, such as cortisol (stress) levels, galvanic skin response, heart-rate variability, and neurotransmitter levels (dopamine, serotonin, oxytocin), but also robust neurometrics, such as brain signals and eye-tracking data formerly obtainable only in the lab. These data might then be mapped to predict an individual's mental state and behavior. Objective metrics could prompt growth in many scientific fields, with a new understanding of cognition and emotion and the possibility of addressing problems like consciousness.

The potential application of objective metrics and quantitative definitions to mental processes also raises the issue of neural-data privacy rights, especially if technological advancement means easier detection of others' states (imagine a ceiling-based reader detecting the states of a whole roomful of people). Biometric data is sensitive as a health-data privacy concern and neural data even more so. There's something special about the brain which is deeply personal, and the tendency is toward strong privacy in this area. For example, many people are willing to share their personal genomic data but not their Alzheimer's disease risk profile. Neural-data privacy rights could be a worry but are overall an invitation for progress. Tools are already in development that could help: diverse research ecosystems, tiered user participation models, and a response to malicious hacking.

Given the high potential value of neural data to science, it's likely that privacy models will be negotiated to move forward with projects. There could be pressure to achieve scale quickly, both in the amount and types of data collected and the validity and utility of the data (still at issue in some areas of personalized genomics). Raw data streams need to be linked to neurophysiological states. Already an ecosystem of open and closed research models is evolving to accommodate different configurations of those conducting and participating in research. One means of realizing scale is through crowd sourcing, both for data acquisition and analysis. This could be particularly true here as low-cost tools for neural-data monitoring become available to consumers and interested individuals contribute their information to an open data commons. Different levels of privacy preferences are accommodated, as a small percentage of those comfortable sharing their data opt in to create a valuable public good usable by all. Even more than has happened in genomics (but not in longitudinal phenotypic data), open-access data could become a norm in neural studies.

Perhaps not initially, but in a later mainstream future for neural data, we might have a granular tiered permissioning system for enhanced privacy and security. A familiar example is the access tiers (family, friends) in such social networks as Facebook and Google Plus. With neural data, we could have similar (and greater) specificity—for example, allowing professional colleagues into certain neural data streams at certain times of day. However, there may be limitations related to a current lack of understanding of neural data streams generally and how signals may be transmitted, processed, and controlled.

The malicious hacking of neural data streams is a potential problem. Issues could arise in both hacking external data streams and devices (like any other data security breach) and hacking communication going back into the human. The latter is too far ahead for useful speculation, but the precedent could be that of spam, malware, and computer viruses. These are "Red Queen" problems, where perpetrators and responders compete in lockstep, effectively running to stay in place, often innovating incrementally to temporarily outcompete the other. Malicious neural-data-stream hacking will likely not occur in a vacuum; we can expect unfortunate side effects, and we'll need responses analogous to antivirus software.

Rather than being an inhibitory worry, the area of neural-data privacy rights invites us to advance to a new node in societal progress. The potential long-term payoff of the continuous bioneurometric-information climate is significant. Objective metrics data collection and its permissioned broadcast might greatly improve both knowledge of the self and our ability to understand and collaborate with others. As personalized genomics has helped destigmatize health issues, neural data could help destigmatize mental health and behavioral issues, especially by letting us infer the issues of the many from the data of the few. Improved interactions with the self and others could free us to focus on higher levels of capacity, spending less time, emotion, and cognitive load on evolutionary-relic communications problems while transitioning to a truly advanced society.

CAN THEY READ MY BRAIN?

STANISLAS DEHAENE

Neuroscientist, experimental cognitive psychologist, Collège de France, Paris; author, Reading in the Brain: The Science and Evolution of a Human Invention

Like many other neuroscientists, I receive my weekly dose of bizarre e-mails. My correspondents seem to have a good reason to worry, though: They think their brain is being tapped. Thanks to new "neurophonic" technologies, someone is monitoring their mind. They can't think a thought without its being immediately broadcast to Google, the CIA, news agencies worldwide . . . or their spouses.

This is a paranoid worry, to be sure. Or is it? Neuroscience is making giant strides, and you don't have to be schizophrenic to wonder whether it will ever crack the lockbox of your mind. Will there be a time, perhaps in the near future, when your innermost feelings and intimate memories will be laid bare for others to scroll through? I believe that the answer is a cautious no—at least for a while.

Brain-imaging technologies are no doubt powerful. More than fifteen years ago, at the dawn of functional magnetic resonance imaging, I was already marveling at the fact that we could detect a single motor action: Any time a person clicked a button with the left or right hand, we could see the corresponding motor cortex being activated, and we could tell with more than 98-percent accuracy which hand the person had used. We could also tell which language the scanned person spoke. In response to spoken sentences in French, English, Hindi, or Japanese, brain activation would either invade a large swath of the left hemisphere, including Broca's area, or stay within the confines of the auditory cortex—a sure sign that the person did or did not understand what was being said. Recently we also managed to tell whether someone had learned to read a given script simply by monitoring the activation of the "visual word form area," a brain region that holds our knowledge of legal letter strings.

Whenever I lectured on this research, I insisted on our methods' limitations. Action and language are macrocodes of the brain, I explained. They mobilize gigantic cortical networks that lie centimeters apart and are therefore easily resolved by our coarse brain-imagers. Most of our fine-grained thoughts, however, are encrypted in a microcode of submillimeter neuronal-activity patterns. The neural configurations that distinguish my thought of a giraffe from my thought of an elephant are minuscule, unique to my brain, and intermingled in the same brain regions. Therefore they will forever escape decoding, at least by noninvasive imaging methods.

In 2008, Tom Mitchell's beautiful *Science* paper proved me partly wrong.[i] His research showed that snapshots of state-of-the-art functional MRI contained a lot of information about specific thoughts. When a person thought of different words or pictures, the brain-activity patterns they evoked differed so much that a machine-learning algorithm could tell them apart much better than would be expected by chance. Strikingly, many of these patterns were macroscopic, and they were even similar in different people's brains. This is because when we think of a word, we do not merely activate a small set of neurons in the temporal lobes that serves as an internal pointer to its meaning. The activation also spreads to distant sensory and motor cortices that encode each word's concrete network of associations. In all of us, the verb "kick" activates the foot region of the motor cortex, "banana" evokes a smell and a color, and so on. These associations and their cortical patterns are so predictable that even new, untrained words can be identified by their brain signature.

Why is such brain decoding an interesting challenge for neuroscientists? It is, above all, a proof that we understand enough about the brain to partially decrypt it. For instance, we now know enough about number sense to tell exactly where in the brain the knowledge of a number is encrypted. And, sure enough, when Evelyn Eger, in my lab, captured high-resolution MRI images of this parietal-lobe region, she could tell whether the scanned person had viewed two, four, six, or eight dots, or even the corresponding Arabic digits.[j]

Similarly, in 2006, with Bertrand Thirion, we tested the theory that the visual areas of the cortex act as an internal visual blackboard where mental images get projected. Indeed, by measuring their activity, we managed to decode the rough shape of what a person had seen, and even of what she had imagined in her mind's eye, in full darkness.[k] Jack Gallant, at Berkeley, later improved this technique to the point of decoding entire movies from the traces they evoke in the cortex. His reconstruction of the coarse contents of a film, as deduced by monitoring the spectator's brain, was an instant YouTube hit.

Why, then, do I refuse to worry that the CIA could harness these techniques to monitor my thoughts? Because many limitations still hamper their practical application in everyday circumstances. First of all, they require a ten-ton superconducting MR magnet filled with liquid helium—an unlikely addition to airport security portals. Furthermore, functional MRI works only with a cooperative volunteer who stays perfectly still and attends to the protocol; total immobility is a must. Even a millimeter of head motion, especially if it occurs in tight correlation with the scanning protocol, can ruin a brain scan. In the unlikely event that you are scanned against your will, rolling your eyes rhythmically or moving your head ever so slightly in sync with the stimuli may suffice to prevent detection. In the case of an electroencephalogram, clenching your teeth will go a long way. And systematically thinking of something else will, of course, disrupt the decoding.

Finally, there are limitations arising from the nature of the neural code. MRI samples brain activity on a coarse spatial scale and in an indirect manner. Every millimeter-sized pixel in a brain scan averages over the activity of hundreds of thousands of neurons. Yet the precise neural code that contains our detailed thoughts presumably lies in the fast timing of individual spikes from thousands of intermingled neurons—microscopic events we cannot see without opening the skull. In truth, even if we did, the exact way in which thoughts are encoded still escapes us. Crucially, neuroscience is lacking even the inkling of a theory as to how the complex combinatorial ideas afforded by the syntax of language are encrypted in neural networks. Until we do, we have very little chance of decoding nested thoughts such as "I think that X . . . ," "My neighbor thinks that X . . . ," "I used to believe that X . . ." "He thinks that I think that X . . . ," "It is not true that X . . .," and so on.

There's no guarantee, of course, that these problems will not be solved—next week or next century, perhaps using electronic implants or miniaturized electromagnetic recording devices. Should we worry then? Millions of people will rejoice instead. They are the many patients with brain lesions whose lives may soon change thanks to brain technologies. In a motivated patient, decoding the intention to move an arm is far from impossible, and it may allow a quadriplegic to regain his or her autonomy, for instance by controlling a computer mouse or a robotic arm. My laboratory is currently working on an EEG-based device that decrypts the residual brain activity of patients in a coma or vegetative state and helps doctors decide whether consciousness is present or will soon return. Such valuable medical applications are the future of brain imaging, not the devilish sci-fi devices that we wrongly worry about.

LOSING COMPLETENESS

ANTON ZEILINGER
Physicist, University of Vienna; scientific director, Institute of Quantum Optics & Quantum Information, Austrian Academy of Sciences; author, Dance of the Photons: From Einstein to Quantum Teleportation

What I worry most about is that we are increasingly losing the formal and informal bridges between different intellectual, mental, and humanistic approaches to seeing the world.

Consider Europe in the first third of the 20th century. Vienna at that time was a hotspot for art, science, literature, music, psychology, and many other disciplines. Johannes Brahms, for example, gave music lessons to the Wittgenstein family, and the Vienna Circle of logical positivism, created by mathematicians and philosophers, gave all of us a new way to look at some of the most fundamental questions.

Another example is Erwin Schrödinger, the founder of wave mechanics. He writes in his autobiographical notes of how he nearly became professor of physics in Czernowitz in the Bukowina, today's Ukraine. There he would have had to teach physics to engineers, and that, he writes, would have given him a lot of spare time to devote to philosophy.

In today's world, all these activities—scientific, artistic, whatever—have been compartmentalized to an unprecedented degree. There are fewer and fewer people able to bridge the many gaps. Fields of expertise and activity become narrower and narrower; new gaps open all the time. Part of the cause is certainly the growth of the Internet, which typically provides immediate answers to small questions; the narrower the question, the better the answer. Deep analysis is an endeavor that by its very essence is entirely different from browsing the Web.

I worry that this trend—this narrowing—will continue. And I worry that in the end we will lose significant parts of our cultural heritage and therefore our very identity as humans.

C. P. SNOW'S TWO CULTURES AND THE NATURE-NURTURE DEBATE

SIMON BARON-COHEN
Psychologist, Autism Research Centre, Cambridge University; author, The Science of Evil: On Empathy and the Origins of Cruelty

More than fifty years have passed since C. P. Snow gave the Rede Lecture in the Senate House at Cambridge University. It was May 7, 1959, when he aired his worry that the majority of his colleagues in the humanities were scientifically illiterate and the majority of his colleagues who were scientists were disinterested in literature. His worry was that two cultures had emerged and were less and less able to understand each other. By way of graphic illustration, Snow argued that scientists would struggle to read a Charles Dickens novel and most humanities professors would be unable to state the second law of thermodynamics. "So the great edifice of modern physics goes up," he declared, "and the majority of the cleverest people in the Western world have about as much insight into it as their neolithic ancestors would have had."

Snow was by training a scientist, who turned his hand to writing novels, exemplifying that rare breed of person who attempts to straddle both cultures. In 1962, Cambridge professor of literature F. R. Leavis scathingly wrote of Snow's lack of ability as a novelist, in an effort to rubbish his "two cultures" argument. Leavis's attack was rightly dismissed as *ad hominem*. But was Snow correct?

If he was, then given the remarkable rate of progress in science over the last fifty years, the gulf between these two cultures may have widened. On the other hand, through the efforts of John Brockman and other literary agents and publishers who have encouraged scientists to communicate to the wider public, creating the so-called third culture, science is now very accessible to nonscientists. So has the gap between Snow's two cultures become wider or narrower?

I think the answer is both. The gap has narrowed thanks to wonderful books like Steven Pinker's *The Language Instinct.* It should now be virtually impossible for a linguist to see language as just a product of culture instead of also a product of our genes. Pinker's book epitomizes what the third culture should be like, illustrating the complex interplay between biology and culture in producing human behavior. Scientists find the idea of a biology/culture interaction unsurprising, almost truistic. As a psychologist, I can think of few if any examples of human behavior that are entirely the result of culture, and I assume that most people interested in human behavior adopt the same moderate position of acknowledging a biology/environment interaction. To be a hard-core biological determinist or a hard-core social determinist seems extreme.

I studied Human Sciences in Oxford in the 1970s, which some people joked was right in the middle of the Banbury Road, the Department of Social Anthropology being on one side and the Department of Biological Anthropology on the other. The Human Sciences students felt like bilingual children, who could not only switch between the two cultures when appropriate but automatically thought about topics in a multidisciplinary way, even if their academic "parents" in each department rarely crossed the road to learn about the other's culture. I would like to think we've come a long way and that there is now rich interchange between disciplines, at least in the study of human behavior.

But I worry that the gap between C. P. Snow's two cultures has in some areas remained as wide as ever and may even have widened. By way of illustration, consider the field of sex differences in the mind. My own view is that research into sex differences teaches us two things: First, one cannot infer what kind of mind a person will have purely on the basis of their gender, since an individual may be typical or atypical of their gender. Indeed, to do so would be stereotyping and sexist. Second, where one finds sex differences *on average* when comparing groups of males and groups of females, these differences are likely to reflect a mix of causal factors, from parenting styles and peer-group influences to the amount of testosterone the fetus produces in the womb and the effects of sex-linked genes.

However, even today one still finds academics claiming that there are no universal sex differences in, for example, language—on the grounds that any sex differences in language and communication are either culture-specific or do not replicate. Such claims effectively reduce sex differences in language to peculiarities of a particular culture or a particular experiment, thereby needing no reference to biology. While I would agree that the similarities in men's and women's conversational styles are greater than are the differences, when it comes to *children's* language acquisition my reading of the evidence is that the differences, on average, between boys' and girls' language development are nontrivial and likely to be universal. Here are just two pieces of experimental evidence.

First, girls typically show faster growth in vocabulary size than boys. This is seen in a large Russian study of 550 girls and 487 boys, aged 18-36 months, mirroring patterns found in a different culture, England. Second, boys' rate of stuttering and other speech problems is at least twice as high as that of girls. This is revealed in an even larger data set from the National Survey of Children's Health, which sampled more than 91,000 children aged three to fourteen across the United States, including children of different ethnic backgrounds. Social determinists might want to take the data from such studies and try to explain it purely in terms of postnatal experience, but since mutations in genes (such as GNPTAB and CNTNAP2) have been associated with stuttering and language impairment, it's likely that individual differences in typical language development (including typical sex differences) will also turn out to have a partly genetic basis.

No one denies the important role that experience and learning play in language development. What worries me is that the debate about gender differences *still* seems to polarize nature vs. nurture, with some in the social sciences and humanities arguing that biology plays no role at all, apparently unaware of the scientific evidence to the contrary. If he were still alive today, C. P. Snow might despair, as I do, that despite efforts to communicate the science to a wider public, the field of sex differences remains a domain where the two cultures are separated by a deep chasm.

THE UNAVOIDABLE INTRUSION OF SOCIOPOLITICAL FORCES INTO SCIENCE

NICHOLAS A. CHRISTAKIS
Physician, social scientist, Harvard University; coauthor (with James Fowler), Connected: The Surprising Power of Our Social Networks and How They Shape Our Lives

I am not really worried about the bad things science might do to society. I'm happy to put aside fears about nuclear power, genetically modified foods, or even the publication of viral genetic sequences. Instead, I'm much more worried about the bad things society might do to science, and I think we should all be.

Lately, I've been seeing a lot of alarming and nonbeneficial interventions by government in science—laws prohibiting the Centers for Disease Control and Prevention from analyzing the epidemiology of gun violence; laws requiring the teaching of "intelligent design" to our children; laws governing stem-cell science; laws affecting climate science. Politicians and pundits take to the Internet and television, like priests of old, to denounce science in ways that sound almost medieval to my ears.

Of course, ever since the Inquisition summoned Galileo, social, political, and religious considerations have affected the conduct of science. Actually, ever since Archimedes was paid to design weapons of war, social considerations have had this effect. Indeed, we can trace such effects as far back as we have records. It has always been the case that social forces have shaped scientific inquiry—what we study, how we study, why we study, who gets to study.

Nevertheless, though science has always been "socially constructed," this fact has been explicitly characterized only in the last few decades. We can now understand how scientists of centuries past could hold views (and even make "objective" observations) that were not only clearly wrong but also clearly driven by ideology or culture. Views ranging from the phrenological causes of crime to the medical diagnosis of escaped slaves as suffering from the disease of "drapetomania" have all been given a scientific gloss.

But we should be really worried about this age-old and unavoidable intrusion of sociopolitical forces at the present historical moment, because our health, security, and wealth depend so much on progress in science and in ways not appreciated widely enough. How we shape science affects our collective well-being. The key driver of economic growth may well be our cumulative growth in knowledge. Science and invention make us richer, and the pace of scientific discovery has been surging for the last 200 years, coincident with, or causally in advance of, economic growth.

Hence, we should pay attention when science (and science education) becomes a plaything of politicians, or when scientists come to be seen like any other interest group (the same as farmers or bankers) rather than as something altogether different. We should be worried that political interference in, and even antipathy to, science harms us all. I'm not suggesting that scientists should be cut off from society, free from moral scruples or collective oversight. But seeing science as arbitrary or threatening and scientists as (merely) self-interested—and using these excuses to restrict scientific inquiry or distort scientific findings—is dangerous.

A discomfiting irony here is that more government support and societal oversight are needed, given changes in how science is done in the 21st century. Long gone are the days when lone scientists with modest resources (Newton, Darwin, Curie, Cavendish, Cox) could make major discoveries. Doing the best science increasingly requires large-scale resources and interdisciplinary teams. We have no choice but to rely on broad social support and the public purse if we are to do our work. So political considerations are unavoidable. We don't want—nor could we even have—unfettered, unexamined, unchecked, or insular scientific inquiry.

So there's no way out of the conundrum. To those who want it both ways—who want the public to support science but also to butt out—my answer is yes, I do, too, very much. We should all be worried if politicians and the public lose sight of the fact that scientific inquiry is a public good.

THE GROWING GAP BETWEEN THE SCIENTIFIC ELITE AND THE VAST "SCIENTIFICALLY CHALLENGED" MAJORITY

LEO M. CHALUPA

Neurobiologist; ophthalmologist; vice president for research, The George Washington University

On a cross-country flight, I once found myself sitting next to a successful attorney who remembered attending a lecture I had given on brain research about ten years earlier at a private club in San Francisco. During our conversation, he asked me if I was still trying to figure out how the brain works. When I indicated that I was still doing research in this field, he seemed surprised, because he thought that after ten years of effort this would be all figured out.

At that moment, it struck me that this highly educated man had no understanding of how science works. He was scientifically ignorant, and the degree he had received at a leading research university before entering law school failed to educate him in the most basic tenet of the scientific process—that research is a never-ending quest.

Now, consider the vast factual ignorance that has repeatedly been documented in recent years. Most people do not believe in evolution; a substantial proportion believe that Earth is only a few thousand years old; many think that vaccines do more harm than good; and (a particularly troubling one for those of us in the brain sciences) it is commonly thought that the brain is a muscle.

Contrast this dismal state of affairs with another personal experience. This involved the hosting of the Siemens High School Science Competition by The George Washington University, where I serve as vice president of research. The finalists in this competition, selected from high schools around the country, presented the results of their research projects. These far surpassed anything I could have possibly accomplished when I was a student at New York's elite Stuyvesant High School in the 1960s. Indeed, as several of the professor judges remarked, the research of these high school students was on a par with that of graduate students or even postdoctoral fellows.

So here is the crux of my worry: the growing gap between the small minority of Americans who are part of the scientific elite and the vast majority who are, to put it kindly, scientifically challenged. This is a worry on several different levels. For one thing, support and funding for research is vitally dependent on informed voters, and even more so on scientifically literate elected representatives. Moreover, as our world faces progressively more challenges (think climate change), how we deal with these complex issues is dependent on an understanding of science and the scientific method.

It is also worrisome that, for the most part, our educational institutions from grade school through college do not teach science the way scientists actually do science. Far too often, science courses involve memorization of a vast array of seemingly unrelated "facts," many of which are of questionable validity. We must do far better, and we need to do so now. Students at all levels should be introduced to science by asking questions and designing experiments to test specific hypotheses. Every town and city should have a hands-on children's science museum, and every professional scientific organization should have a community outreach program. Some of this is already being done, but the gap between the scientifically informed and uninformed continues to grow. So I continue to worry and will do so until we come up with a realistic plan to reverse this troubling trend.

PRESENT-ISM

NOGA ARIKHA
Historian of ideas; chair, Critical Studies, Paris College of Art; author, Passions and Tempers: A History of the Humours

I worry about the prospect of collective amnesia.

Whereas access to information has never been as universal as it is now thanks to the Internet, the total sum of knowledge of anything beyond the present seems to be dwindling among those who came of age with it. Anything before 1945 (if then) is a messy, remote landscape; the centuries melt into each other in an insignificant magma. Famous names are flickers on a screen, their dates irrelevant, their epochs dusty. Everything is equalized.

The stunning historical blankness displayed by students the world over when they arrive at college can be explained. For one thing, confusion has set in about what should be taught. The canon—in any country—is now considered, by many who are supposed to teach it, an obsolete, "imperialist" weapon to be shunned, not an expandable, variable set of works that have passed the test of time and should begin the learning process. Chronology is moot: Instead, students can pick general themes, analyze perspectives and interpretations. History certainly needs to remain a thoughtful enterprise—one would not want it to turn back into a list of dates, monarchs, and battles. But because the laudable emphasis on questioning has taken the place of teaching questionable facts, temporal continuities have been lost, and students have few guides other than the Internet, which has become their reference, their ersatz library.

So there's the rub of technology. Those who come of age today are challenged by the inventions from which their elders reap benefit: Facebook disperses everyone's attention but captures theirs in a particularly intense way. Wikipedia is a useful shortcut everyone can use as a starting point for proper research, but students use it as if it constituted all research. Without a background in leafing through books, learning in the old analog way how to gauge relevance, hierarchy, accuracy, and cross-references, students arrive at college unable to know where to begin their education and so where to place themselves within the wider world; they are confused about what constitutes the beginning of knowledge. A few do learn how and what to ask and may find the guidance to delve into the world's histories, acquiring a sense of perspective and of what it means to know that one doesn't know. But for a majority, the gaps will persist into maturity, unidentified and unplugged.

Certainly worries emerge whenever technology changes the mode of cultural transmission. Though there might be a correlation between the acceleration of technological transformations and the speed at which the past recedes, one should not forget that the fear of forgetting was strong when printing became common. If forgetfulness has increased, it is not because of new technology per se but despite it: It is because curricular fashions are mimicking Internet entropy rather than providing the centripetal force needed to turn the young into its informed users. As it is, ours is an age of information glut, not deep knowledge.

The very way in which science is practiced, for instance, rewards short-term memory combined with a sense that the present exists for the sake of the future. Ten-year-old scientific papers are now ancient; after all, over a million new papers are published each year. As a result, some groundbreaking work from the 1920s in, say, zoology, lies forgotten and is produced in new labs as if it had never been done before. Almost everything is archived, but nothing can be found, unless one knows to look for it. Many may be reinventing the wheel, unaware that the historical permafrost is full of treasures.

The same applies to the longer history of science, the history of art, the history of philosophy, and political and economic history. (Some of our present woes arguably arise out of the historical myopia of economists.) References to early modernity have become a highly specialized affair, hard to place in chronology. For many, and not only for students, history is of the modern, not of what precedes the modern, when science was magic, all art was the same, and all politics autocratic. History is not so much simplified as disappeared. There are pockets of awareness, pre-modern eras considered fascinating by a general public and picked up by Hollywood; for the West, these would include, say, ancient Egypt, Greco-Roman antiquity, the "Middle Ages," the Renaissance, the Moghuls in India, the American Civil War, the French Revolution. But each is viewed as a self-contained epoch—easily skippable episodes in the potentially entertaining soap opera or costume drama that is the history of the world. And so, deeper questions of historiography cannot even begin to be addressed.

Of course, history is still researched, written, read by enough individuals that the discipline as such survives. The sum of historical knowledge has always been held by a small number of educated people at any given time, and this has not changed. But our world is geared at keeping up with a furiously paced present, with no time for the complex past; and the fact that a very large number of literate people with unprecedented access to advanced education and scanned sources has no sense of what the world was like only yesterday points to the possibility of eventually arriving at a state of collective amnesia. We risk remaining stuck within a culture where everyone ignores the sundry causal connections that make the present what it is, happy to focus on today's increasing complexity—as if a blank slate favored creativity and innovation.

There is a way out: by integrating the teaching of history within the curricula of all subjects, using whatever digital or other means we have to redirect attention to slow reading and old sources. Otherwise we will be condemned to live without perspective, robbed of the wisdom and experience with which to build for the future, confined by the arrogance of our presentism to repeating history without noticing it.

DO WE UNDERSTAND THE DYNAMICS OF OUR EMERGING GLOBAL CULTURE?

KIRSTEN BOMBLIES

Assistant professor of organismic & evolutionary biology, Harvard University

Humanity is building a global superculture. We know it's dynamic and exciting, but do we understand what we're getting into? Like local cultures, this global one exists in a parallel world of information only loosely tied to physical substrates or individual minds. It has a life of its own, and we don't fully understand its evolutionary potential. As a result, we probably don't adequately appreciate the possible hazards associated with it. What unanticipated emergent properties might a globalized culture have?

Luckily there are informative parallels between biological and cultural evolution. Cultural systems, like biological systems, change over time. Heritable units of culture—counterparts of genes in biology—are difficult to define, but we can observe information propagating and mutating: for example, as the packets that Richard Dawkins popularized as "memes." Unlike genes, ideas (or memes) can be transmitted worldwide in an instant and potentially reach huge audiences, so the evolution of cultures can hurtle on at a rate far exceeding that of our biological evolution. The mutation and differential survival of ideas, however, is analogous to the mutation and differential survival of genetic variants, and thus our understanding of cultural evolution can be informed by biological concepts.

A central lesson from biological evolution is that those genetic variants that propagate best come to numerical domination. Though they often achieve this by benefiting their carriers, some variants are so good at replicating and propagating that they can increase even if they harm their carriers. Some ideas spread similarly. The mass propagation of a factually incorrect, divisive, or misleading idea is not uncommon. Is this a disease state or a beneficial and natural part of the dialog in a free society? A complication in answering this is that one person's dangerous idea may be another's revelation. Is there a recognizable distinction between a harmless or beneficial flow of ideas and a malignant state?

Despite complications in defining cultural disease, it's clear that selfish ideas can propagate—we have seen it in our lifetime. I mean "selfish" here in an evolutionary sense: as something that's good at increasing numerically without conferring any benefit to its host. Sociology and psychology can address the question of why some ideas spread and others don't, but it seems inevitable that selfish replicants will at some time proliferate in virtually any evolving system. In every ecosystem, there are parasites; in almost every genome, there are selfish elements; in every society, there are cheats. Thus we should probably consider what detrimental properties might propagate in a global culture and how their troublesomeness scales with culture size and complexity. Is there, or could there be, a cultural immune system—for example, is the system sufficiently self-policing, with the infusion of counterviews from diverse people? We need to know whether something can go seriously or systematically wrong with a global culture, and how to recognize and fix it if it does. I hope we can ensure that its flaws are minor and that it becomes a generally positive force for humanity.

WE WORRY TOO MUCH ABOUT FICTIONAL VIOLENCE

JONATHAN GOTTSCHALL
Adjunct professor of English, Washington & Jefferson College; author, The Storytelling Animal: How Stories Make Us Human

In the wake of a mass shooting, we feel a desperate need to know "why?" so we can get to "how"—how can we keep this from happening again? When someone shoots up a school, a mall, or an office, people on the left usually blame lax gun laws and recommend making them stricter. On the right, people are apt to blame cultural factors—violent video games and films that have sickened the culture, glorifying wanton violence and desensitizing young people to its effects—and see loose gun regulation not as a cause of the massacres but as our best defense against them.

But this idea of blaming the media is an oldie and a baddy. First, speaking practically, where would we draw the line? If we managed to ban trigger-happy games like *Doom*, *Call of Duty*, and *Halo*, what would we do about violent films like *Saving Private Ryan* or equally gory classics like Homer's *Iliad*? Should we edit the killings out of Shakespeare's plays?

Second, the evidence that violent media promotes violent behavior is pretty shaky. Violence is a great—perhaps *the* great—staple of the entertainment economy. As a society, we guzzle huge amounts of fake violence in television shows, novels, films, and video games. And yet a determined fifty-year search for real-world consequences of fictive violence hasn't found conclusive evidence of a causal linkage. Some researchers argue that the more we consume violent media, the more likely we are to behave aggressively in the real world. Other researchers disagree, picking studies apart on methodological grounds and pointing out that many hundreds of millions of people watch violent television and play violent games without developing the slightest urge to kill. As scientists like Steven Pinker point out, we consume violent entertainment at a greater rate than we ever have before, yet we've never been at lower risk of a bloody demise. The more we've consumed, the more peaceable and law-abiding we've become.

Has the consumption of violence in the media actually helped reduce criminal violence? The notion isn't as perverse as it may at first seem. Critics of media violence envision scenarios that have us vicariously reveling in wanton savagery. But the messages found in most video games are strongly prosocial. Adventure-style video games almost always insert players into scenarios wherein they play the role of a hero bravely confronting the forces of chaos and destruction. When you play a video game, you aren't training to be a spree shooter; you are training to be the good guy who races to place himself between evil and its victims.

The same goes for traditional fiction formats—film, television, and novels. Virtually without exception, when the villain of a story kills, his violence is condemned. When the hero kills, he does so righteously. Fiction preaches that violence is acceptable only under defined circumstances—to protect the good and the weak from the bad and the strong. Some games, such as *Grand Theft Auto*, seem to glorify and reward bad behavior (although in a semi-satirical spirit), but those games are the notorious exceptions that prove the general rule. What Stephen King says of horror stories in his book *Danse Macabre* broadly applies to all forms of imaginary violence:

> [T]he horror story, beneath its fangs and fright wig, is really as conservative as an Illinois Republican in a three-piece pinstriped suit. . . . [I]ts main purpose is to reaffirm the virtues of the norm by showing us what awful things happen to people who venture into taboo lands. Within the framework of most horror tales we find a moral code so strong it would make a Puritan smile.

So how should we respond to mass-shooting tragedies? First, we must resist the reflex to find and torch a scapegoat, whether in the entertainment industry or the gun lobby. Even if we could keep unstable people from consuming imaginary violence, they could still find plenty of inspiration from the nightly news, history, holy Scripture, or their own fevered dreams. And even if we were able to pass laws that kept guns out of the hands of bad men (a tall order in a country with an estimated 300 million guns in private circulation), how would we keep them from killing with improvised explosives or by plowing SUVs into crowds?

Second, hard as it is, we must struggle to keep these acts of terrorism in perspective. We are all in much more danger from a simple traffic accident on the way to a school, a theatre, a political rally than we are from gunfire once we get there. We would save many more lives by, say, focusing on highway safety than by putting guns in the hands of "good people" or rekindling culture wars over the role of guns and entertainment in American life. It's a cliché, but true: When we overreact, the terrorist wins. When we overreact, we give the sicko meme of the mass shooting the attention it needs to thrive.

So here's what shouldn't worry us: fictional violence.

Here's what should: the way our very understandable pain and fear lead us to respond ineffectively to real violence.

A WORLD OF CASCADING CRISES

PETER SCHWARTZ
Futurist, business strategist; senior vice president for global government relations & strategic planning, Salesforce.com; author, Inevitable Surprises

Hardly a day passes when we do not hear of some major crisis in the world, one already unfolding or a new one just beginning, and they never seem to end. We are now living in a world of perpetual crisis and the high anxiety it produces. Crises are not new. Human societies have experienced natural disasters, wars, famines, revolutions, political collapse, plagues, depressions, and more. But two conditions are new. First, the interconnection of the world's many systems often lead one crisis to cascade into the next like falling dominoes. Second, because of global communications and new media, we are all far more aware of more of the crises than ever before.

One of the best examples of the first is the recent financial crisis that began in the sub-prime mortgage market, when fear outweighed hope and cascaded through our banking system, our economy, triggering the Euro crisis and slowing growth in China. And the monetary measures we took to save the economy at the height of the crisis may yet trigger an inflationary crisis, when growth starts to recover and capacity fails to catch up with new demand.

The public's view of violence is a good example of the second new dynamic, that of heightened awareness. Steven Pinker has elegantly shown us that the real threat of violence in most of our lives has diminished dramatically. Yet most of us believe, because of the constant drumbeat of reporting on violence, that the threat is far greater than it actually is. The absence of children playing on suburban streets is a sign of how scared parents are of the threat of kidnapping, which actually remains very small.

So if some of the issue is an increase in crises driven by interconnection, and some merely a perceptual issue, why should we be worried? Because living in a world of high anxiety often leads us to do the wrong things. We adopt short-term and local solutions rather than taking a systemic and long-term view. Does anyone believe that continuing financial innovation will not lead to another financial crisis in the near future? And yet the structural issues are not even being discussed. Our huge overreactions to 9/11 and to drug use have led to a perpetual security state from which we cannot return. Vast numbers of people are incarcerated at a huge cost for something that should not even be a crime. And we all pay the Bin Laden tax when we put up with the costs and disruptions of airport security. And of course our political systems play into this with a vengeance, because they are notoriously poor at systemic and long-term solutions, and those perverse instincts are amped up by the public sense of perpetual crisis. There has been an inevitable loss of faith in institutions' ability to get ahead of the curve and tamp down the trembling state of anxiety the world now seems unable to shake off.

There is no apparent way back from this sense of teetering on the precipice. We cannot disconnect the world's functional systems and we cannot disconnect our awareness from the world's information systems. This is our new reality: high anxiety driven by cascading crises, real and imagined.

WHO GETS TO PLAY IN THE SCIENCE BALLPARK

STEPHON H. ALEXANDER
Ernest Everett Just 1907 Professor of Natural Sciences & associate professor of physics & astronomy, Dartmouth College

I am worried about who gets to be a player in the science game—and who is left out. As I was growing up in New York, I was always encouraged and rewarded for my involvement in jazz. Fellow musicians, elders, and contemporaries, while demanding excellence, remained inclusive; everyone was given a chance to solo, and if you were good enough you could get called back up.

Unfortunately, with the exception of a few enlightened individuals, professional inclusion has not been the collective experience of underrepresented groups in certain scientific disciplines. (This goes beyond ethnicity and includes underrepresented people who also "think differently.") Given the U.S. population trends, some argue that for us to be competitive in STEM-related fields we'll need to extend the scientific enterprise to the increasing Latino and black populations. I worry that while funding structures tend to address this issue at the K-12 and college levels, there has been little serious discussion about recruiting and promoting persons of color in the academy. Much research has shown that a lack of role models for these underrepresented groups negatively affects excellence and retention.

The issue of lack of inclusivity in science goes beyond the usual discourse surrounding affirmative action. Exceptionally talented individuals, regardless of their background, should not be made to feel that an opportunity was bestowed on them because of a policy instead of their abilities. The scientific community needs to go beyond tolerance of difference, to the *genuine appreciation* of difference, including those differences that make us feel uncomfortable. Especially an appreciation of those who see the world differently from us and think differently from us.

I understand that the issue of racial inclusivity is sensitive and loaded with complicated sociological and political overtones. I do think it is an elephant in the room that needs to be dealt with by us as a scientific community. We need to be honest and open. As the demographics of the United States (and the world at large) change, do we care to prepare our academies, at all levels, to share resources and influence with the currently underrepresented. In the spirit of my friend Brian Eno's contribution, I have taken the challenge to bring up something that may make people uncomfortable—because it is more productive to do so than to be polite.

AN EXPLODING NUMBER OF NEW ILLEGAL DRUGS

THOMAS METZINGER
Philosopher, Johannes Gutenberg-Universität Mainz; author, The Ego Tunnel

I have been predicting this for a number of years. But now it is really happening, and at a truly amazing, historically unprecedented speed. The number of untested but freely available psychoactive substances is dramatically rising. In the European Union in 2011, new drugs were detected at the rate of roughly one per week according to the EMCDDA–Europol annual report on new psychoactive substances released on April 26, 2012. A total of forty-nine new psychoactive substances were officially notified for the first time in 2011, via the EU early-warning system. This represents the largest number of substances ever reported in a single year, up from forty-one reported in 2010 and twenty-four in 2009. All in all, this means 164 new drugs since 2005—but also that the annual record has now been broken for three years in a row.

All of the new compounds reported in 2011 were synthetic. They are cooked up in underground labs; and increasingly organized crime develops the market and imports them—for example, from China. Next to nothing is known about their pharmacology, toxicology, or general safety; almost all these substances have never been tested *in vivo* or in animal models. This makes the situation difficult for medical staff at psychiatric emergency units, with kids tripping on substances whose names doctors have never heard of, even during their university education; the substances did not yet exist while the doctors were being trained.

I sometimes naïvely assumed that in the end most of these substances would turn out to be fairly benign, but even for "spice drugs" (synthetic cannabinoids sprayed on herbal smoking mixtures, often marketed as "legal highs"), deaths and serious heart problems have now been reported in young healthy adolescents.

Last October, a twenty-one-year-old in New Orleans died from taking just a single drop of 25-I (25I-NBOMe), a new synthetic hallucinogenic drug. In the United States, at least three more deaths from this new phenethylamine (originally discovered in Germany) have been reported in Minnesota and North Carolina, the victims being seventeen, eighteen, and twenty-five years old.

Everybody knows about acute adverse reactions, psychotic effects, addiction—we do have some cultural experience. But what about long-term effects, such as early-onset cognitive decline, say—a somewhat steeper slope than in normal aging? Unexpected carcinogenic agents? It will take us a long time to discover and verify such effects.

Does anybody remember thalidomide? The situation is now completely out of hand. And it is a complicated situation: We might ban (or attempt to ban) early generations of substances with low health-hazard profiles, creating a psychopharmacological arms race, leading to replacement by ever newer and potentially more dangerous compounds.

In 2011, the list of substances registered was dominated by synthetic cannabinoids (twenty-three new compounds on the market) and synthetic cathinones (eight new molecules). They now represent the two largest drug groups monitored by the European early warning system and make up around two-thirds of the new drugs reported that year; but, of course, this may change soon. The War on Drugs has failed, but the process of scientific discovery is going on. The number of substances in the illegal market will increase and perhaps altogether new categories of drugs will emerge—plus the corresponding "neurophenomenological state-classes," as I like to call them.

What are the main causal factors that have brought about this development?

First, it's a combination of scientific progress and human nature: Fundamental research—for example, in neuropharmacology—simply moves on, and new knowledge and newly developed technologies meet age-old human desires for spiritual self-transcendence, hedonism, and recreation. Plus greed. For me, new psychoactive substances are a paradigm example of how neurotechnology turns into consciousness technology (I like to call it "phenotechnology"). Low-level nuts-and-bolts molecular neuropharmacology creates not only new forms of subjective experience but also high-level effects for society as a whole: invisible risks, new industries, emerging markets.

The second causal factor is the Internet. An Internet snapshot conducted in January 2012 showed 693 online shops offering at least one psychoactive substance or product, up from 314 in January 2011 and 170 in January 2010. Recipes for many of these new illegal substances, as well as first-person reports of the phenomenology associated with different dosages, are available on the Internet—easily accessed by the alternative psychotherapist in California, the unemployed chemistry professor in the Ukraine, or the Chinese mafia. And now the sheer increase in the speed with which new drugs appear on the market makes all established procedures obsolete.

But there is at least one deeper reason for the situation we will face in the decades to come: Half a century ago, the Western world deliberately opted for a culture of denial and repression.

In the sixties, we witnessed a semisynthetic psychedelic drug of the ergoline family spread around the world, with millions of new users and a new generation of underground chemists. The appearance of LSD was a historically new situation. We might perhaps have had a chance to deal with the challenge then, rationally, based on ethical arguments. The really difficult, and most important, task would have been to weigh the psychological and social risks against a general principle of liberty and the intrinsic value of the experiences resulting from one or another altered brain state. But we decided to push it underground and look the other way. Half a century later, a globalized and scientifically well-informed underground chemistry strikes back. More kids are going to die. It would be starry-eyed optimism to think we can still control the situation.

HISTORY AND CONTINGENCY

PAUL KEDROSKY
Investor; contributing editor, Bloomberg & Partner, SK Ventures

How many calls to a typical U.S. fire department are actually about fires? Less than 20 percent. If fire departments aren't getting calls about fires, what are they mostly getting calls about? They are getting calls about medical emergencies, traffic accidents, and, yes, cats in trees, but they are rarely being called about fires. They are, in other words, organizations that despite their name deal mostly with everything but fires.

Why, then, are they called fire departments? Because of history. Cities used to be built out of pre-combustion materials—wood straight from the forest, for example—but they are now mostly built of post-combustion materials—steel, concrete, and other materials that have passed through flame. Fire departments were created when fighting fires was a more urgent urban need, and now their name lives on, a reminder of their host cities' combustible past.

Everywhere you look, you see fire departments: not literally fire departments, but organizations, technologies, institutions, and countries that, like fire departments, are beyond their "past due" date or weirdly vestigial yet remain widespread and worryingly important.

One of my favorite examples comes from the siting of cities. Many U.S. river cities are where they are because of portages—the carrying of boats and cargo around impassable rapids. This meant, many times, overnight stays, which led to hotels, entertainment, and, eventually, local industry, at first devoted to shipping but then broader. Now, however, those portage cities are prisoners of history, sitting along rivers that no longer matter for their economy, meanwhile struggling with seasonal floods and complex geographies antithetical to development—all because a few early travelers, using transportation technologies that no longer matter, had to portage around a few rapids. To put it plainly, if we rebooted right now, most of these cities would be situated almost anywhere else first.

This isn't just about cities or fire departments. This is about history, paths, luck, and "installed base" effects. Think of incandescent bulbs. Or universities (or tenure). Paper money. The post office. These are all examples of organizations or technologies that persist largely for historical reasons, not because they remain the best solution to the problem for which they were created. They are often obstacles to much better solutions.

Obviously this list will get longer in the near future. Perhaps multilane freeways join the list, right behind the internal combustion engine. Or increasingly costly and dysfunctional public markets. Malls as online commerce casualties. Or even venture capitalists in the age of AngelList and Kickstarter. How about geography-based citizenship? All of these seem vaguely ossified, like they're in the way—even if most people aren't noticing, yet.

But this is not a list-making game. This is not some Up with Technology exercise where we congratulate ourselves at how fast things are changing. This is the reverse. History increasingly traps us, creating paths—and endowments and costs, both in time and money—that must be traveled before we can change directions, however desirable those new directions might seem. History—the path by which we got here, and the endowments and effluvia it has left us—is an increasingly large weight on our progress. Our built environment is an installed base, like an ancient computer operating system that holds back progress because compatibility gives such an immense advantage.

The writer William Gibson once famously said, "The future is already here—it's just not very evenly distributed." I worry more that the past is here—it's just so evenly distributed that we can't get to the future.

UNKNOWN UNKNOWNS

GARY MARCUS

Professor of psychology; director, NYU Center for Language & Music; author, Guitar Zero: The New Musician and the Science of Learning

There are known knowns and known unknowns, but what we should be worried about most is the unknown unknowns. Not because they are the most serious risks we face but because psychology tells us that unclear risks in the distant future are the risks we're less likely to take seriously enough.

At least four distinct psychological mechanisms are at work. First, we are moved more by vivid information than by abstract information (even when the abstract information should in principle dominate). Second, we discount the future, rushing for the dollar now as opposed to the two dollars we could have a year later if we waited. Third, the focusing illusion (itself perhaps driven by the more general phenomenon of priming) tends to make us dwell on our most immediate problems even if more serious problems loom in the background. Fourth, we have a tendency to believe in a just world, in which nature naturally rights itself.

These four mechanisms likely derive from different sources, some stemming from systems that govern motivation (future discounting), others from systems that mediate pleasure (belief in a just world), others from the structure of our memory (the focusing illusion, and the bias from vividness). Whatever their source, the four together create a potent psychological drive for us to underweight distant future risks we cannot fully envision.

Climate change is a case in point. In 1975, the Columbia University geochemist Wallace S. Broecker published an important and prescient article in *Science* called "Climatic Change: Are We on the Brink of a Pronounced Global Warming?" but his worries were ignored for decades, in part because many people presumed, fallaciously, that nature would somehow automatically set itself right. (And in keeping with our tendency to draw inference primarily from vivid information, a well-crafted feature film on climate change played a significant role in gathering public attention, arguably far more so than the original *Science* article.)

Oxford philosopher Nick Bostrom has pointed out that the three greatest unknowns we should worry about are biotechnology, nanotechnology, and the rise of machines more intelligent than human beings. Each sounds like science fiction and has in fact been portrayed in science fiction, but each poses genuine threats. Bostrom posits "existential risks"—possible, if unlikely, calamities that would wipe out our entire species, much as an asteroid appears to have extinguished the dinosaurs. Importantly, many of these risks, in his judgment, exceed the existential risk of other concerns that occupy a considerably greater share of public attention. Climate change may be more likely, and certainly is more vivid, but is less apt to lead to the extinction of the human species (even though it could conceivably kill a significant fraction).

The truth is, we simply don't know enough about the potential biotechnology, nanotechonology, or future iterations of artificial intelligence to calculate what their risks are. Compelling arguments have been made that in principle any of the three could lead to human extinction. These risks may prove manageable, but I don't think we can manage them if we don't take them seriously. In the long run, biotech, nanotech, and AI are probably significantly more likely to help the species, by increasing productivity and limiting disease, than they are to destroy it. But we need to invest more in figuring out exactly what the risks are and preparing for them. Right now, the United States spends more than $2.5 billion dollars a year studying climate change but (by my informal reckoning) less than 1 percent of that total studying the risk of biotech, nanotech, and AI.

What we really should be worried about is that we are not quite doing enough to prepare for the unknown.

DIGITAL TATS

JUAN ENRIQUEZ
Managing director, Excel Venture Management; author,
As the Future Catches You: How Genomics & Other Forces Are Changing Your Life, Work, Health & Wealth

"How else can one threaten other than with death? The interesting, the original thing, would be to threaten someone with immortality." Jorge Luis Borges' response to the thugs in the Argentine Junta combines guts and fear in equal measure. He did not so much fear torture in this world as much as a life continually reexamined and reappraised throughout the ages.

We should all heed his warning, as we are all rapidly becoming immortal through electronic tattooing. As we all post more and more information on who we are, what we do, like, dislike, think, and say, a big data portrait emerges that gets ever harder to lose, modify, erase. In a big data world, we don't just leave bread crumbs behind, we voluntarily and involuntarily leave giant, detailed pointillist portraits of our every day.

It used to be only royalty, presidents, megastars, and superstar athletes who had every aspect of their daily lives followed, analyzed, scrutinized, criticized, dissected. Now ubiquitous cameras, sensors, tolls, RFIDs, credit cards, clicks, friends, and trolls describe, parse, and analyze our lives minute by minute, day by day, month by month. Habits, hatreds, opinions, desires are recorded and will be visible for a long, long time. What you wore and ate, with whom and where, what you said and did, where you slept. . . . There is more than enough data and inside dirt on almost all of us to enable a weekly *Page 6* embarrassment, a *People* magazine hero profile, and a detailed biography, bitter and sweet.

Few understand how recently the shift to ubiquitous and permanent recording occurred. High-definition video was expensive and cumbersome; that's why you don't see hundreds of September 11th real-time videos on YouTube. Handheld, hi-def movie cameras, ubiquitous and cheap today, were not widely available in 2001. It used to be that most of the street cameras were posted and hosted by gated communities, store owners, police or traffic authorities. Not only have these security cameras gotten more and more effective (a single DC camera issued 116,734 tickets in less than two years) but video cameras of all types became almost disposably inexpensive. So cheap, high-def cameras and sensors have spread like bedbugs. We are surrounded by thousands of cameras operated by endless security groups, cabs, home monitors, someone's phone, babysitters, beach boardwalk promoters, surfers, marketers, "citizen" newsmakers, weather bugs, and dozens of other folks with faint excuses to film permanently. It's not just the filming itself that's the game changer; rather, it is the almost negligible cost of archiving. This used to be expensive enough that tape was simply overrecorded after a few minutes or hours. Now all is kept. We now see and know one another in ways previously unimaginable. (Just browse the hundreds of rapidly breeding reality-TV channels and millions of YouTube posts.)

Every time we blog, charge, debit, Tweet, Facebook, Google, Amazon, YouTube, LinkIn, Meetup, Foursquare, Yelp, Wikipedia, we leave electronic nuggets, some more visible than others, of who we are, whom we are with, and what we like or are interested in. In a sense, we electronically tattoo ourselves, our preferences, our lives, in a far more comprehensive and nuanced way than any inked skin. Trivially easy to apply, seemingly innocuous, initially painless, these frequent electronic tattoos are long-lasting and someday will portray you as an almost saint or a serious sinner or both. Electronic tattoos are trivially easy to copy, reproduce, spread, store. They will long outlive your body. So in a sense they begin to fulfill Borges' greatest fear: They begin to make one immortal.

Tattoos are serious commitments. Every parent knows this. Often kids do not. Once inked, a tattoo is a lifelong commitment to a culture, cause, person, passion, hatred, or love. One cannot belong to certain tribes or gangs without a public billboard that promises till death do us part. Sometimes, once inked, there is no hiding, and it is hard, if not impossible, to change sides. Whether on a beach, bed, classroom, job, or cell, every pair of eyes judges and thinks it knows who you are, what you believe in, whom you play with. Tattoos publicly advertise fidelity, dedication, love, hate, and stupidity. That which, after a few tequilas in Vegas, may have seemed a symbol of never-ending romance can become a source of bitter conversation on the future honeymoon. Now multiply that embarrassment, disclosure, past history a thousandfold using electronic tattoos. (Type "high schoolers" into Google and the first four search hints to appear are "making out," "grinding," "in bikinis," and "kissing." Eventually, do you really want to see the detailed video of future grandma without gramps?)

Immortality and radical transparency have consequences. We are at immediate risk of having the world know what we do, what we did, ever more available for scrutiny by our peers, rivals, bosses, lovers, family, admirers, as well as random strangers. SceneTap cameras let you know, at any given minute, how many people are in dozens of Boston bars, the male/female ratio and average age. This will soon seem quaint. Facial-recognition technologies enable you, with a 90-percent-plus accuracy, to use a smartphone to identify someone standing at a bar. Add name recognition and location to your phone and likely you can quickly access a series of details on whether that cute person over there has any criminal convictions, where she lives, what her property is worth, how big a mortgage, Yelp preferences, Google profile, Facebook status, Tweets, reunion notes. . . .

These overall records of our lives will be visible, accessible, hard to erase for a long, long time. It's so painless, so easy, cheap, and trivial to add yet another digital tat to our already very colorful and data-filled electronic skin that we rarely think about what it might mean, teach, say about us in the long term. So we have covered our bodies, our images, ourselves in far more detail than even the most tattooed person on the planet.

And even if you acted impeccably according to today's norms and customs, electronic immortality still presents enormous challenges. Religions, ethics, customs, and likes change. What is currently deemed simply distasteful may be judged quite differently by successor generations: Knowing vaguely that great Greek philosophers sexually consorted with young boys is different from having direct access to lurid tapes during history class. Detailed visuals on how the Founding Fathers treated their slaves might seriously affect their credibility. Today's genetics has already uncovered and detailed Thomas Jefferson's peccadilloes. Were today's technologies and detailed histories available a few centuries ago, likely we would have far fewer recognized saints. No doubt some things we do as a matter of course today will seem rude, barbaric, or even criminal to classrooms full of future high schoolers trolling through huge databases of lives.

FAST KNOWLEDGE

NICHOLAS HUMPHREY
Darwin College, University of Cambridge; author, Soul Dust: The Magic of Consciousness

In 1867, Dostoevsky, on a visit to the cathedral in Basel, was transfixed by Holbein's portrait of the dead Christ. According to his wife's diary, he climbed on a chair to look more closely at the painting and remained for a good half hour, taking in every detail. Two years later, in his novel *The Idiot*, he was able to give an uncannily accurate description of the painting, as if he had photographed it in his mind.

In 2013, of course we would not need to go to such trouble. We could snap the painting on an iPhone, and summon it up later on the Retinex display. Actually, we wouldn't have to visit Basel. We could check it out on Google Images back home. And while we're at it, we would only have to type in "Dostoevsky + Holbein" to find this anecdote confirmed at a dozen sites.

One-touch knowledge can be an enormous boon. Yet I worry that in raising us all to a plane of unprecedented genius, it is creating a drearily level playing field. When each of us can learn so much, so easily, in the same way as everybody else, we are in danger of becoming mere knowledge tourists, hopping from attraction to attraction at 30,000 feet without respecting the ground that lies between. One-step travel is a boon as well. But when everyone finds themselves going to the same places, when it's the arrival and not the journey that matters, when nothing memorable happens along on the way, I worry that we end up, despite our extraordinary range of experience, with less to say.

In the old days, when, like Dostoevsky, we had to work at it, the learning we accumulated, however eccentrically, was both valued and valenced. The landscape of our knowledge had mountains and valleys, flat sands and gushing geysers. Some parts of this territory we had explored for ourselves, others we were guided to and paid for, others we might have chanced on by luck. But however we came by them, we were proud to know the things we did. They were the presents we brought to the table of intellectual debate. In argument we could reveal them when and if we chose, we could put everything out front, we could pretend ignorance, and play hard to get.

We should worry that this dimension of individual intelligence is disappearing, and with it that flirtatiousness that leads to the marriage of ideas. Soon no one will be more or less knowledgeable than anyone else. But it will be knowledge without shading to it, and, like the universal beauty that comes from cosmetic surgery, it will not turn anyone on.

SYSTEMATIC THINKING ABOUT HOW WE PACKAGE OUR WORRIES

MARY CATHERINE BATESON
Cultural anthropologist; professor emerita, George Mason University; Visiting scholar, Sloan Center on Aging & Work, Boston College; author, Composing a Further Life: The Age of Active Wisdom

This seems to be a moment for some systematic thinking about how we package our worries. Most people seem to respond only if they have the ability to visualize a danger and empathize with the victims. This suggests the need for:

(1) A realistic time frame. Knowledgeable people expected an eventual collapse of the Shah's regime in Iran but did nothing, because there was no pending date. In contrast, many prepared for Y2K because the time frame was so specific.
(2) A specific concern for those who will be harmed. If a danger lies beyond my lifetime, it may seem significant if it threatens my grandchildren. People empathize more easily with polar bears and whales than with honeybees and bats. They care more about natural disasters in countries they have visited.
(3) A sense of what the proximate danger is, which is often a byproduct or side effect of what is talked about. "Global warming" was a bad description of a danger, because it sounded comfy, and even "climate change" sounds fairly neutral. Extreme weather conditions causing humanitarian disasters get more attention. Regional warfare for access to resources

(oil) or arable land may be more salient than a few degrees of temperature change or rising ocean surfaces. It is entirely possible that global warming will lead to nuclear war as a side effect, but that's not where our concern needs to be.

There is a need for research on the social psychology of fear and anxiety, which is undoubtedly going to be different from what we know about the individual psychology of fear and anxiety. For instance, it seems probable that a sense of chronic threat is a permanent element in some populations. What has replaced fear of the "red menace" in Americans, and how has this replacement affected attitudes to immigration or to the deficit? How essential is the state of constant threat to Israeli social solidarity? How consciously has the U.S. government manipulated the fear of terrorism in U.S. politics? How does the fear of "stranger danger" displace the fear of domestic violence? Arguably, humans may require a certain amount of worry to function effectively, whether this worry is fear of hell or fear of the neighbors. If this is the case, it might be safer to focus on worrying about the Red Sox winning the pennant.

WORRYING ABOUT STUPID

ROGER SCHANK
Artificial intelligence theorist; cognitive scientist; CEO of SocraticArts; author, Teaching Minds: How Cognitive Science Can Save Our Schools

I am worried about stupid. It is all around us. When the Congress debates an issue, both sides appear to be wrong. Worse, our representatives can't seem to make a reasoned argument. Candidate after candidate in the last election said things completely unsupportable by the evidence. Supporters of these candidates couldn't usually explain in any coherent way what they were really in favor of.

Or consider experiences in customer service, where it is clear that the person with whom you are speaking is reading from a script and incapable of deviating from the script because they really don't know what they're talking about.

Our high school graduates are very good at test taking. Story after story comes out about how cheating occurs even in the best schools and even involves teachers. No one asks what students are capable of doing after they graduate, because no one cares. We just worry about what they've memorized. I worry about whether they can think. Talk to a recent graduate and see if they can.

I worry that young people don't talk any more. LOL and OMG do not a conversation make. It is conversations that challenge beliefs, not emoticons.

I worry that since no one thinks they need to think, the news has become a mouthpiece for views that can be easily parroted by their listeners. Challenging beliefs is not part of the function of the news anymore.

I worry that we have stopped asking "Why?" and stopped having to provide answers. We say we need more math and science in school and don't ask why. We say we need more soldiers and more arms and don't ask why. We say a prescription drug works miracles, but we fail to ask about what we really know about what else it does.

We glorify stupid on TV shows, showing what dumb things people do, so that we can all laugh at them. We glorify people who can sing well but not those who can think well. We create TV show after TV show demonstrating that acting badly will make you rich and famous. The fact that most exchanges on these shows are done without thought and without the need to back up opinions with evidence seems to bother no one.

I worry that behind this glorification of stupidity and the refusal to think hard about real issues are big corporations who make a great deal of money on this. The people who sell prescription drugs don't want people asking how the drug works or how the clinical trials went or what harm the drug might do. The people who debate spending cuts don't want people asking why they never talk about cuts in defense or cuts in the gigantic sums we give to other countries. The people who run corporations that profit on education don't want anyone to ask whether there aren't already many unemployed science and math PhDs; they just want to make more tests and sell more test prep. The people who run universities don't want anyone to ask whether a college education is really necessary for the majority of the population, or even well done at the majority of our colleges. The people who run news organizations have an agenda, and it isn't creating good thinkers who understand what's going on in the world.

So I'm worried that people can't think, can't reason from evidence, and don't even know what would constitute evidence. People don't know how to ask the right questions, much less answer them. And I'm worried that no one, with the exception of some very good teachers who are much unappreciated, is trying to teach anyone to think.

I'm worried that we will, as a society, continue to make decisions that are dumber and dumber, and that there will be no one around smart enough to recognize these bad decisions or able to do anything about them.

THE CULTURAL AND COGNITIVE CONSEQUENCES OF ELECTRONICS

LUCA DE BIASE

President, Fondazione Ahref; scientific director, Digital Accademia; visiting lecturer in new media & journalism, IULM University, Milan

Sixty years ago, the Italian poet Giuseppe Ungaretti wrote about the cultural future of electronics. As a young poet before World War I, Ungaretti had been briefly fascinated by the Futurist movement, and he had strong feelings about the "poetic" rhythm of mechanics and the aesthetics in the technical discoveries. But after World War II, at the age of sixty-five, he was worried. In the past, he wrote, technology followed human imagination, but in the future the enormous set of engineering achievements, led by electronics, was going to go faster than human imagination; thus humanity could end up thinking as machines, losing the ability to feel, love, fear.

Lately, many scholars have been giving more than a thought to these matters. It has been a generative debate, touching on everything from the way search engines change our memorization strategies to the persuasive effects of interface design and even to the way some open platforms enhance our tendency to cooperate instead of compete. But after all that thinking, we are paradoxically both bored and worried. Innovation in digital technologies moves at such a fast pace that we still need more research about the cultural and cognitive consequences of electronics. For this research to succeed, it must be focused on long-term changes, which means defining the problem in a more holistic and less deterministic way, as Ungaretti suggested.

How can we think about the way we think, without letting our minds be trapped in the means we use to communicate? The mediasphere is a sort of environment, in which most information and knowledge live and develop. And the "information ecosystem" is a generative metaphor to make sense of such a media environment. This metaphor helps mediologists understand the coevolution of ideas and platforms. It works if one wants to stress the importance of diversity in the mediasphere. And it leads to approaching the media as a complex system.

But as the information-ecosystem metaphor grows more popular, it also generates a set of inconvenient analogies. Thus we start to look for an "ecology of information," and we worry about long-term tendencies. This leads us to define specific problems about the risks of monocultures, the possible development of info-polluting agents, the existence of unsustainable media practices. This approach can be useful if we don't let ourselves misinterpret the metaphor. Science, economics, politics, entertainment, and even social relations grow in the mediasphere. And ideologies, misinformation, and superstition also develop in the information ecosystem. We cannot think of info-pollution in terms of "bad content," because nobody can define it as such, just as nobody can define as "bad" any life-form in an ecosystem. Info-pollution is not about the content; it is more about the process. It is about the preservation of a cultural equilibrium.

What exactly are we looking for, then?

We are looking for ways to let our imagination be free, to breathe new ideas, to think in a way not explained only by the logic and incentives of the mediasphere. How can this be done?

Poetry is a kind of research that can help. Digital humanities are a path to enhance our ability to think differently. We need to develop an epistemology of information.

Knowledge is meant to set us free, provided that we preserve our ability to decide what is important, independent of what is defined as important by the platform we use. Freedom of expression is not only the quantity of different ideas that circulate (the wealth of which has never been as rich as it is today). Freedom of expression is also about the decision making that lets us choose ideas that are better than others, to improve our ability to live together.

We should be worried about how we go about finding the wisdom to allow us to navigate developments, as we begin to improve our ability to cheaply print human tissue, grow synthetic brains, have robots take care of our aging parents, let the Internet educate our children. Ungaretti would have thought, maybe, that what is required is only a digitally aware name for ethics, aesthetics, and poetry.

WHAT WE LEARN FROM FIREFIGHTERS: HOW FAT ARE THE FAT TAILS?

NASSIM NICHOLAS TALEB

Distinguished Professor of Risk Engineering, NYU Polytechnic Institute; author, Antifragile: Things That Gain from Disorder

Eight years ago, I showed, using 20 million pieces of data from socioeconomic variables (about all the data that was available at the time), that current tools in economics and econometrics don't work whenever there is an exposure to a large set of deviations, or "Black Swans." There was a gigantic mammoth in the middle of the classroom. Simply, one observation in 10,000—that is, one day in forty years—can explain the bulk of the "kurtosis," a measure of what we call "fat tails," that is, how much the distribution under consideration departs from the standard Gaussian, or the role of remote events in determining the total properties. For the U.S. stock market, a single day, the crash of 1987, determined 80 percent of the kurtosis. The same problem is found with interest and exchange rates, commodities, and other variables. The problem is not just that the data had "fat tails"—something people knew but sort of wanted to forget; it was that we would never be able to determine how "fat" the tails were. Never.

The implication is that those tools used in economics that are based on squaring variables (more technically, the Euclidian, or L-2 norm), such as standard deviation, variance, correlation, regression, or value-at-risk—the kind of stuff you find in textbooks—are not valid *scientifically* (except in some rare cases where the variable is bounded). The so-called "p values" you find in studies have no meaning with economic and financial variables. Even the more sophisticated techniques of stochastic calculus used in mathematical finance do not work in economics except in selected pockets.

The results of most papers in economics based on these standard statistical methods—the kind of stuff people learn in statistics class—are thus not expected to replicate, and they effectively don't. Further, these tools invite foolish risk-taking. Neither do alternative techniques yield reliable measures of rare events, except that we can tell if a remote event is underpriced, without assigning an exact value.

The Evidence

The story took a depressing turn, as follows: I put together this evidence—in addition to *a-priori* mathematical derivations showing the impossibility of some statistical claims—as a companion for *The Black Swan*. The papers sat for years on the Web, were posted on *Edge* (ironically the *Edge* posting took place only a few hours before the announcement of the bankruptcy of Lehman Brothers). They were downloaded tens of thousands of times on SSRN (the Social Science Research Network). For good measure, a technical version was published in a peer-reviewed statistical journal.

I thought that the story had ended there and that people would pay attention to the evidence; after all, I played by the exact rules of scientific revelation, communication, and transmission of evidence. Nothing happened. To make things worse, I sold millions of copies of *The Black Swan* and nothing happened, so it cannot be that the results were not properly disseminated. I even testified in front of a congressional committee (twice). There was even a model-caused financial crisis, for Baal's sake, and nothing happened. The only counters I received were that I was "repetitive," "egocentric," "arrogant," "angry," or something even more insubstantial, meant to demonize the messenger. Nobody has managed to explain why it is not charlatanism, downright scientifically fraudulent, to use these techniques.

Absence of Skin in the Game

It all became clear when, one day, I received the following message from a firefighter. His point was that he found my ideas on tail risk extremely easy to understand. His question was: How come risk gurus, academics, and financial modelers don't get it?

Well, the answer was right there, staring at me, in the message itself. The fellow, as a firefighter, could not afford to misunderstand risk and statistical properties. He would be directly harmed by his error. In other words, he has skin in the game. And, in addition, he is honorable, risking his life for others, not making others take risks for his sake.

So the root cause of this model fraud has to be absence of skin in the game, combined with too much money and power at stake. Had the modelers and predictors been harmed by their own mistakes, they would have exited the gene pool—or raised their level of morality. Someone else (society) pays the price of the mistakes. Clearly, the academic profession consists in playing a game, pleasing the editors of "prestigious" journals, or being "highly cited." When confronted, they offer the nihilistic fallacy that "We've got to start somewhere"—which could justify using astrology as a basis for science. And the business is unbelievably circular: A "successful PhD program" is one that has "good results" in the "job market" for academic positions. I was told bluntly at a certain business school where I refused to teach risk models and "modern portfolio theory" that my mission as a professor was to help students get jobs. I find all of this highly immoral—immoral to create harm for profit. *Primum non nocere.*

Only a rule of skin in the game—that is, direct harm from one's errors—can puncture the game aspect of such research and establish some form of contact with reality.

LAMPLIGHT PROBABILITIES

BART KOSKO
Information scientist & professor of electrical engineering & law, University of Southern California; author, Noise

We should worry that much of our science and technology still uses just five main models of probability, even though there are more probability models than there are real numbers. I call these *lamplight* probabilities. The adjective refers to the old joke about the drunk who lost his keys somewhere in the dark and looks for them under the streetlamp because that's where the light is.

The five lamplight probabilities do explain a lot of the observed world. They have simple closed-form definitions. So they are easy to teach. Open any book on probability and there they are. We have proved lots of theorems about them. And they generalize to many other simple closed-form probabilities that also seem to explain a lot of the world and find their way into real-world models ranging from finance to communications to nuclear engineering.

But how accurate are they? How well do they match fact rather than just simplify the hard task of picking a good random model of the world?

Their use in practice seldom comes with a statistical hypothesis test that can give some objective measure of how well the assumption of a lamplight probability fits the data at hand. And each model has express technical conditions that the data must satisfy. Yet the data routinely violate these conditions in practice.

Everyone knows the first lamplight probability: the normal bell curve. Indeed, most people think that the normal bell curve is *the* bell curve. But there are whole families of bell curves with thicker tails that explain a wide range of otherwise unlikely "rare" or "black swan" events, depending on how thick the tails are. And that is still assuming that the bell is regular or symmetric. You just don't find these bell curves in most textbooks.

There is also a simple test for such thin-tailed "normality" in time-series data such as price histories or sampled human speech: All higher-order cumulants of the process must be zero. A cumulant is a special average of the process. Looking at the higher-order cumulants of an alleged normal process routinely leads to the same finding: They're not all zero. So the process cannot be normal. Yet under the lamplight we go ahead and assume the process is normal anyway—especially since so many other researchers do the same in similar circumstances. That can lead to severely underestimating the occurrence of rare events, such as loan defaults. That's just what happened in the engineering models of the recent financial panic, when financial engineers found a way to impose the normal curve on complex correlated derivatives.

The second and third lamplight probabilities are the Poisson and exponential probability models. Poisson probabilities model random counting events, such as the number of hits on an Internet site or the number of cars merging onto a freeway or the number of raindrops hitting a sidewalk. Exponential probabilities model how long it takes for the next Poisson event to happen—how long it takes until the next customer walks through the door or the next raindrop hits the pavement. This generalizes into how long you have to wait for the next ten Internet hits or the next ten raindrops. Modern queuing theory rests on these two lamplight probabilities. It's all about waiting times for Poisson arrivals at queues. And so the Internet itself rests on these two lamplight models.

But Poisson models have an Achilles heel: Their averages must equal their variances (spreads about their means). Again, this routinely fails to hold in practice. Exponential models have a similar problem; their variances must equal the square of their means. This is a fine relationship that also rarely holds exactly in practice, holding only to some fuzzy degree in most cases. Whether the approximation is a good enough is a judgment call—and one that the lamplight makes a lot easier to make.

The fourth lamplight probability is the uniform probability model. Everyone also knows the uniform model, because it is the special case where all outcomes are equally likely. It is just what the layman thinks of as doing something "at random," such as drawing straws or grabbing a numbered Ping-Pong ball from a Bingo hopper. But straws come in varying lengths and thicknesses and so their draw probabilities may not be exactly equal. It gets harder and harder in practice to produce equally likely outcomes as the number of outcomes increase. It is even a theoretical fact that one cannot draw an integer at random from the set of integers, because of the nature of infinity. So it helps to appeal to common practice under the lamplight and simply assume that the outcomes are all equally likely.

The fifth and final lamplight probability is the binomial probability model. It describes the canonical random metaphor of flipping a coin. The binomial model requires binary outcomes such as heads or tails and further requires independent trials or flips. The probability of getting heads must also stay the same from flip to flip. This seems simple enough. But it can be hard to accept that the next flip of a fair coin is just as likely to be heads as it is to be tails when the preceding three independent coin-flip outcomes have all been heads.

Even the initiated scratch their heads at how the binomial

behaves. Consider a fair penny. Fairness here means that the penny is equally likely to come up heads or tails when flipped (and hence the fourth lamplight probability describes this elementary outcome). So the probability of heads is 1/2. Now flip the penny several times. Then answer this question: Are you more likely to get three heads in six coin flips or are you more likely to get three heads in just five coin flips? The correct answer is neither. The probability of getting three heads in both cases is exactly 5/16. That is hardly intuitive, but it comes straight out of counting up all the possible outcomes.

Lamplight probabilities have proven an especially tight restriction on modern Bayesian inference. That's disappointing, given both the explosion in modern Bayesian computing and the widespread view that learning itself is a form of using Bayes' Theorem to update one's belief given fresh data or evidence. Bayes' Theorem shows how to compute the probability of a logical converse. It shows how to adjust the probability of lung cancer given a biopsy, if we know what the raw probability of lung cancer is and the probability that we would observe such a biopsy in a patient if the patient in fact has lung cancer. But almost all Bayesian models restrict these probabilities to not just the well-known lamplight probabilities. They further restrict them so that they satisfy a very restrictive "conjugacy" relationship. The result has put much of modern Bayesian computing into a straitjacket of its own design.

Conjugacy is tempting. Suppose the cancer probability is normal. Suppose also that the conditional probability of the biopsy, given the cancer, is normal. These two probabilities are conjugate. Then what looks like mathematical magic happens. The desired probability of lung cancer, given the biopsy likelihood, is itself a normal probability curve. And thus normal conjugates with normal to produce a new normal. That lets us take this new normal model as the current estimate of lung cancer and then repeat the process for a new biopsy.

The computer can grind away on hundreds or thousands of such data-based iterations and the result will always be a normal bell curve. But change either of the two input normal curves and in general the result will not be normal. That gives a perverse incentive to stay within the lamplight and keep assuming a thin-tailed normal curve to describe both the data and what we believe. The same thing happens when we try to estimate the probability of heads for a biased coin using the binomial and a beta curve that generalizes the uniform. And a similar conjugacy relation holds between Poisson probabilities and exponentials and their generalizations to gamma probabilities. So that gives three basic conjugacy relations. Most Bayesian applications assume some form of one of these three relations—and almost always for ease of computation or to comply with common practice.

That's no way to run a revolution in probabilistic computing.

THE WORLD AS WE KNOW IT

RICHARD FOREMAN
Playwright & director; founder, the Ontological-Hysteric Theater

The responses to this year's *Edge* Question are concerned with problematic aspects of this world or of current discourse that are, I believe, ultimately noncorrectable through either direct thought or direct action. Time, of course, will alter the parameters within which these problems are situated, and they will be absorbed or transcended by new, evolving parameters that will eventually dissolve them. Even if "the end of the world itself" is threatened, that too will lead to some other state—inconceivable perhaps, but even as a vague allusion we humans will dismiss it as beyond thought and not to be wished for. But that understandable rejection arises from the same psychological base implied in the word "worried," which is the linch pin of this year's Question.

One can say that the "worried" can be thought of as a means of focusing upon a particular problem. But the very act of picking one problem out of the many available constructs the trap into which we all fall the minute we begin thinking about the world. We should in fact be "worried" not just about a single selected problem but about *all possible* problems.

But most important: What does "worried" mean, other than the inevitable fall into human consciousness that focuses the mind—producing inevitably science, politics, and everything else in *the world as we know it.* There seems no responsible alternative; indeed, it is the only historical alternative to the disreputable "blissed out" state of passivity and removal from the real world, as known to us through our rigorously conditioned mechanisms.

Perhaps—but perhaps not. I reference not only suppressed mystical traditions—plus more acceptable philosophers, such as Heidegger and phenomenologically oriented contemporaries—but first and foremost my early collegiate inspiration in art theory, Anton Ehrenzweig's great books *The Hidden Order of Art* and *The Psychoanalysis of Artistic Vision and Hearing*. Ehrenzweig demonstrates how artists in many disciplines, from many different historical periods, operate not out of normal focused vision but out of wide-angle unfocused perception—and thereafter soon discovered similar theses, hidden or not, in other "official" Western thought. But how does theory relate to real-world problems of the sort we should now be worried about? Well, de-focusing on obsessive worrisome problems often leads to the sudden emergence of a solution where previous directed effort had often failed. (Eureka!—Poincaré, etc.)

So what should we be worried about? Perhaps the failure to stop worrying, when stopping can (all by itself, after the proper preparing of the ground with concentration and "worry") lead to sudden vision.

It's tricky, yes. And difficult. And sometimes a frightening risk—giving up everything we "know" when it's "knowing" that gives rise to the uncontrollable virus of worrying. But I suggest that this year's question is a hidden trick. That was not the intention, I feel sure, but I do see it as a trick question for all half-brilliant, half-sleeping human consciousness.

WORRYING—THE MODERN PASSION

JAMES J. O'DONNELL
Classical scholar, University Professor, Georgetown University; author, The Ruin of the Roman Empire

Worrying is a worry. Identifying a serious problem and taking rational action to analyze and mitigate it: excellent behavior. Identifying a serious problem, taking what steps you can to mitigate it, and recognizing that you can't do more: excellent behavior.

But worrying about things is a modern passion. To worry properly, you should worry in groups, large groups; you should identify something that could happen but doesn't have to; you should forget to read Daniel Kahneman's *Thinking, Fast and Slow;* and you should proceed to dwell on the topic, seeking to promote anxiety for yourself and especially for others; you should be very sure not to engage in rational (especially quantitative) analysis or heed the work of those who do; and you should abstain from all action that might make the problem go away.

Your anxiety will in the end go away, because the problem will most likely go away; or perhaps your fear will come true and you'll be in a different place; or else you'll be dead. You will have maximized your unhappiness and stress levels and, with luck, those of others, with nothing to show for it otherwise.

The ancients knew this behavior, of course, and the Greeks called it *deisidaemonia* and the Latins *superstitio*. They used those words to bracket behavior that led to no good. We've borrowed the latter word but applied it narrowly, to activities where anxiety cringes before the imagined divine. The ancients didn't need a broader category, because the ubiquitous divine could be blamed for anything. We haven't got a better word yet to describe and deplore generalized cringing before what our own imaginations show us. We could use one. In the meantime, a broad awareness of the cultures and deeds of people and nations in our time, and all the times before that we know of, can be an excellent remedy for worry and an instigation to rational behavior.

There are a lot of other essays on this typically intriguing *Edge* Question that invoke worry and use the word easily—so should we be worried about that? Here's the test I would use to decide: If what my colleagues in these pages are talking about is evidence-based, especially quantitative, peer-reviewed or peer-reviewable, and has the effect of increasing our understanding and moving people toward appreciation of intelligent actions we can take as individuals or society, I'm not going to call that "worry." And there is a lot of it.

But if it's about the Comet ISON veering off its path and plunging to Earth and there's nothing we can do about it—that's a worry we should worry about.

THE GIFT OF WORRY

ROBERT PROVINE
Psychologist & neuroscientist, University of Maryland; author, Curious Behavior: Yawning, Laughing, Hiccupping, and Beyond

Worry is a kind of thought and memory evolved to give life direction and protect us from danger. Without its nagging whispers, we would be prone to a reckless Panglossian lifestyle marked by drug abuse, unemployment, and bankruptcy. Why not smoke, have that last drink for the road, pollute the water supply, or evade taxes? All will be fine! In our feel-good, smiley-face era, generally free of plague, famine, and war, it's easy to adopt a smug optimism and demean a behavior that has served us well throughout our evolutionary history.

In the popular media, we are constantly reminded of the presumed personal and social benefits of chirpy unbridled optimism and are taught to fear the corrosive effects of pessimism and worry. I speak from experience. As an expert on laughter and humor, I am frequently contacted by reporters seeking quotes for stories they have been commissioned to write about "laughing your way to health," "the power of a positive outlook," and the like. They are uninterested when I report that laughter, like speech, evolved to change the behavior of other people, not to make us healthy, and that any medicinal effects of laughter are secondary. Also unwelcome are my comments about the dark side of laughter in jeers, ridicule, and violence. In the interest of restoring balance, I accept the role of curmudgeon and defend the value of worry.

On the basis of a famous long-running longitudinal study of 1,178 high-IQ boys and girls initiated by Lewis Terman in 1921, Howard Friedman and colleagues found that conscientiousness was the personality trait that best predicted longevity. Contrary to expectation, cheerfulness (optimism and sense of humor) was *inversely* related to longevity. In other words, conscientiousness, a correlate of worry, pays off in a long life. A modest level of worry is usually best, an instance of the U-shaped function familiar in physiological and behavioral science. Too much worry strands us in an agitated state of despair, anxiety, and paranoia; too little leaves us without motivation and direction. Worry contributes life's to-do list, but its relentless prompts are unpleasant, and we work to diminish them by crossing items off the list. The list is constantly fine-tuned and updated. As life's problems are solved, topics of worry are extinguished, or if a dreaded event does not occur or becomes obsolete, we substitute new, more adaptive topics of concern. The bottom line? Stop worrying about worry. It's good for you.

NOTES

**NO SURPRISES FROM THE LHC:
NO WORRIES FOR THEORETICAL PHYSICS**

page 183 a. "Black Holes: Complementarity or Firewalls?" arXiv:1207.3123v2 [hep-th] 22 Aug. 2012.

GLOBAL GRAYING

page 234 b. "Human Population Grows Up," *Scientific American*, September 2005.

THE DECLINE OF THE SCIENTIFIC HERO

page 273 c. "An experimental study of apparent behaviour," *Amer. J. Psychol.* 13 (1944).

page 274 d. Castelli, F., Frith, C., Happé, F., Frith, U., "Autism, Asperger syndrome and brain mechanisms for the attribution of mental states to animated shapes," *Brain* 125(8): 1839-49 (2002).

IS IDIOCRACY LOOMING?

page 345 e. "Cognitive Capitalism: The Effect of Cognitive Ability on Wealth, As Mediated Through Scientific Achievement and Economic Freedom," *Psychol. Sci*, 22:6, 754-63 (2011).

**SHOULD WE WORRY ABOUT BEING ABLE
TO UNDERSTAND EVERYTHING?**

page 387 f. Susan Kruglinski, "When Even Mathematicians Don't Understand the Math," *New York Times*, May 25, 2004.

THE DEMISE OF THE SCHOLAR

page 391 g. Techonomy Conference, Lake Tahoe, California, August 6, 2010.

**ILLUSIONS OF UNDERSTANDING AND
THE LOSS OF INTELLECTUAL HUMILITY**

page 397 h. Marcia C. Linn, et al. "Can Desirable Difficulties Overcome Deceptive Clarity in Scientific Visualizations?" in *Successful*

Remembering and Successful Forgetting: A Festschrift in Honor of Robert A. Bjork, Aaron Benjamin, ed. (New York: Routledge, 2010).

CAN THEY READ MY BRAIN?

page 411 i. "Predicting Human Brain Activity Associated with the Meanings of Nouns," *Science* 320, 1191 (2008).
page 412 j. "Deciphering Cortical Number Coding from Human Brain Activity Patterns," *Curr. Biol.* 19, 1608-15 (2009).
page 412 k. "Inverse retinotopy: Inferring the visual content of images from brain activation patterns," *NeuroImage* 33:4, 1104-16 (2006).

INDEX